LIGNIN BIOCHEMISTRY

Lignin Biochemistry

BY

WALTER J. SCHUBERT

Fordham University
The Bronx, New York

1965

ACADEMIC PRESS   New York and London

ACADEMIC PRESS INC.
111 Fifth Avenue, New York, New York 10003

United Kingdom Edition published by
ACADEMIC PRESS INC. (LONDON) LTD.
Berkeley Square House, London, W.1

LIBRARY OF CONGRESS CATALOG CARD NUMBER: 65-18436

PRINTED IN THE UNITED STATES OF AMERICA

To Tony, Bill,
Chris, and John

Preface

The past two decades have witnessed a phenomenal worldwide surge of research interest in the chemical and technological aspects of lignin. An appreciation of the scope of the scientific productivity in these areas may be gained from a consideration of the two masterful works by F. E. Brauns, "The Chemistry of Lignin," Academic Press, New York, 1952, and "The Chemistry of Lignin: Supplement Volume" (Covering the Literature for the Years 1949–1958), Academic Press, New York, 1960.

This period has also been marked by the gradual supplementation of the classical investigative techniques of organic and physical chemistry in the study of lignin with what may be described as an application of the "biochemical approach" to the lignin problem. This assumed the form of an investigation of the actual biochemical phenomena associated with lignin, as well as an application of biochemical and biophysical methods to the solution of certain more familiar problems encountered in lignin studies, such as its isolation.

There are two fundamental biochemical aspects to the lignin problem: the enzymic mechanism of its biogenesis in living, lignifying plants, and the mechanisms which nature has provided for the gradual removal of the synthesized lignin content of nonliving plant material, such as by the enzymic activities of microorganisms. The burden of this monograph is concerned with these phenomena.

It seemed to the author that these biochemical considerations would be more meaningful, particularly to the nonspecialist, if they were projected against an outline of certain pertinent fundamental aspects of the chemical properties of lignin. Accordingly, the work begins with a very brief review of the structural chemistry of lignin, at least as far as this is known with some degree of certainty.

Since our current knowledge of lignin biogenesis is so intimately related to earlier work on aromatic biosynthesis in microorganisms, a brief review of this work was also felt to be relevant. The principal chapters on lignin biogenesis and degradation follow this review. No attempt was made at a complete evaluation of all relevant literature. The viewpoint throughout is that of the Laboratory of Organic Chemistry and Enzymology at Fordham

University, under the direction of Professor F. F. Nord, where the author has been privileged to study for the past several years.

The author wishes to take this opportunity to express his gratitude to the Officers of Administration of Fordham University for the award of a Faculty Fellowship, and to the National Science Foundation for a grant, which provided the opportunity to prepare the original manuscript of this work. In addition, he would also like to express his indebtedness to Professor Nord for his many valuable suggestions and for his encouragement which was necessary for the completion of the task.

The Bronx, New York   WALTER J. SCHUBERT
April, 1965

Contents

ix

Chapter I • Introduction:
The Chemistry of Lignin

Chemically as well as anatomically, wood is a heterogeneous material. However, despite their complexity and diversity, all species of woods contain three major chemical components. These are cellulose, the hemicelluloses, and lignin. The first two are usually referred to collectively as "holocellulose." All three wood cell wall constituents are complex, high-polymeric compounds. While our knowledge of the detailed constitution of the cellulose component is very extensive, information on the total structure of lignin is still incomplete. Exhaustive reviews on the chemistry of the lignins have appeared (9, 10).

1

After a brief consideration of those fundamentals of the chemical structure of lignin which are generally accepted by most contemporary lignin chemists, the burden of this work will focus on the biochemical aspects of lignin. These include the biosynthetic processes by which the lignin is formed during the life of the growing tree, which is referred to as "lignification," and the processes by which nature has provided for the gradual removal of the lignin component of dead or decaying wood, namely, by the metabolic activities of certain microorganisms that have the capacity of returning the organic structure of lignin back to the carbon cycle.

Before proceeding to a resumé of the early history of investigations into the chemistry of lignin, it should be pointed out that there are also small amounts of inorganic matter present in wood, and somewhat greater amounts of a variety of organic substances which are extractable from wood with water or with certain neutral organic solvents, such as ethyl alcohol, benzene, and diethyl ether.

These materials are referred to as the "extraneous components" of wood, or the "extractives," and are not an integral part of the cellular structure of wood. The quantity of "extractive" components present in wood may vary from as little as 1% to as high as 40–50% (3). The chemical classes of substances that comprise the extractives include almost all known series of organic compounds, i.e., hydrocarbons, alcohols, phenols, ethers, aldehydes, ketones, acids, esters, and lactones. Specific types of these components may include sugars, sugar alcohols, cycloses, fats, fatty acids, terpenes, sterols, quinones, flavones, pigments, tannins, and others.

I. EARLY HISTORY OF LIGNIN CHEMISTRY

For a long time, wood was considered to be a uniform chemical substance. However, with the introduction of the techniques of experimental organic chemistry over 100 years ago, it gradually became possible to separate wood into its individual components. The first serious attempt to study the chemical composition of woody plants was made in 1838 by the French chemist and botanist, Anselme Payen (39). He treated wood alternately with nitric acid and strong alkali and then with ethyl alcohol and ethyl ether, and as a result he obtained a substance that was quite resistant to the action of these reagents and solvents. Payen called this substance "cellulose."

In the process of the isolation of cellulose, Payen observed that it was necessary to remove another substance that had a higher percentage of carbon than did the cellulose. He first referred to this other material as "*la matiere ligneuse véritable*" (the true woody material), but he later designated it as "*la matière incrustante*" (the incrusting material), because he believed that the cellulose in woody plants was surrounded or impregnated with this material. Payen also attempted to isolate this incrusting material as such, but he was not completely successful. However, by modifying his method of extraction, he did obtain a series of preparations which contained varying ratios of cellulose and the incrusting material, and he called these preparations "*lignose*," "*lignon*," "*lignireose*," and "*lignin*." Lignin (from the Latin word for wood, *lignum*) was later adapted in 1857 by Schulze (55) as the name for Payen's "incrusting material," and this name has been retained to the present time.

As a result of the development of the wood-pulping industry, in which huge quantities of lignin were obtained as a by-product in the form of so-called "sulfite" and "black" liquors which were usually discarded, it was natural that attempts would be made to exploit this potentially valuable raw material. But such utilization required some fundamental information on the chemical properties of the material, and so in Sweden, around 1890, Peter Klason (who is often referred to as the "father of lignin chemistry") initiated the first intensive and extensive investigations of lignin chemistry by studying one of its common commercial by-products, namely lignosulfonic acid (27).

CH₂OH
CH
CH

OCH₃
OH

(I)

—C—
—C—
—C—

OCH₃
O

(II)

Among the many contributions of Klason to our knowledge of the chemistry of lignin, one may cite the following: He developed the first method for the quantitative determination of lignin in woody plants; he was the first to isolate lignin by applying sulfuric acid (a procedure that forms the basis for a method of lignin determination still commonly employed); he discovered that if coniferyl alcohol (I) were treated under the same conditions as those used in the sulfite process of pulping wood, this alcohol would be converted into a sulfonic acid possessing many properties similar to those of ligno-sulfonic acid; but most important of all, Klason was the first to suggest that the parent structure of lignin might be a phenylpropane derivative of the coniferyl type (II), a hypothesis which is still widely accepted.

Considering the tremendous amount of wood that goes to the pulp mills annually (in the United States alone, it averages over two billion cubic feet per year), and considering that about 25% of this is obtained as a lignin derivative of one kind or another and is then discarded, it is not surprising that the wood pulping industries are constantly attempting to utilize this substance. Although the wood saccharification industry has not yet found a use for its waste lignin, except as a fuel, the pulp industry has had some success in employing a small part of its waste lignin in binders, adhesives, briquetting agents, synthetic rubber and the like. A small amount of the lignin in sulfite waste liquor, furthermore, is converted into one marketable chemical, namely vanillin, which is used as a synthetic flavoring agent.

One fundamental reason why greater utilization of lignin has not yet been achieved is because our knowledge of its chemical structure is still incomplete. Even though over 100 years have passed since its discovery by Payen, and extensive experimentation has been undertaken on it, we are still unable to write a complete structural formula for this substance.

Due to our incomplete knowledge of the chemistry of lignin, it is very difficult even to define it. There is little doubt that lignin belongs to a class of compounds that are completely different from cellulose and the other polysaccharide carbohydrates. That portion of wood called "lignin" is generally separated from the other components only by the application of strong chemical reagents which undoubtedly have the effect of chemically modifying the lignin. This is one reason why lignin preparations that have been isolated by different methods vary in their chemical compositions quite extensively.

It should be emphasized at the outset that the term "lignin" can no longer be regarded as the designation of one individually defined compound, but rather, it should be thought of more as a collective term for a whole series of similar, very large molecules which are all closely related structurally to one another, perhaps in an analogous way as are certain other natural polymerization products, such as cellulose and starch, or perhaps, the proteins.

Based upon earlier research on lignin, this material was once defined (9, p. 15) as: "that incrusting material of the plant which is built up mainly, if not entirely, of phenylpropane building stones; it carries the major part of the methoxyl content of the wood; it is unhydrolyzable by acids, readily oxidizable, soluble in hot alkali and bisulfite, and readily condenses with phenols and thio compounds." On the basis of some more recent information, the definition of lignin has since been somewhat amplified. Thus, since lignin yields aldehydes when it is treated with nitrobenzene in alkali at 160°, it may be further described as "that wood constituent which, when oxidized with nitrobenzene, yields vanillin (III) in the case of coniferous woods, vanillin and syringaldehyde (IV) in the case of deciduous woods, and *p*-hydroxybenzaldehyde (V), vanillin, and syringaldehyde in the case of

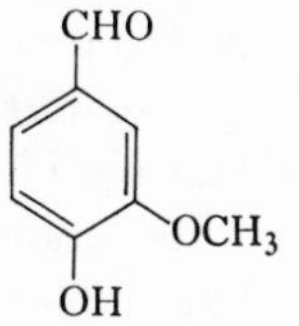

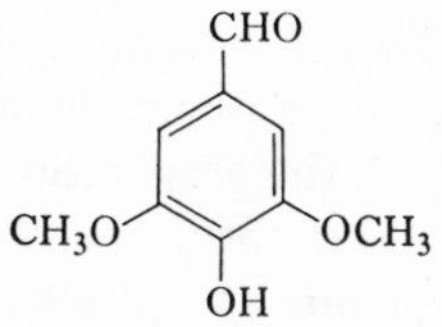

monocotyledons" (10, p. 7). In addition, lignin may be considered as "that plant component which, when refluxed with ethanol in the presence of catalytic amounts of hydrogen chloride, gives a mixture of ethanolysis products— Hibbert's monomers—such as α-ethoxypropioguaiacone (VI), vanillin (III), and vanilloyl methyl ketone (VII) from coniferous woods and, in addition, the corresponding syringyl derivatives from deciduous woods" (10, p. 7).

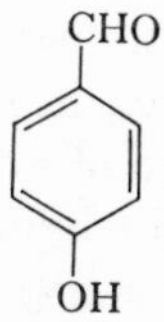

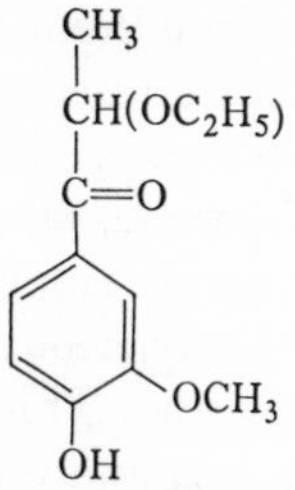

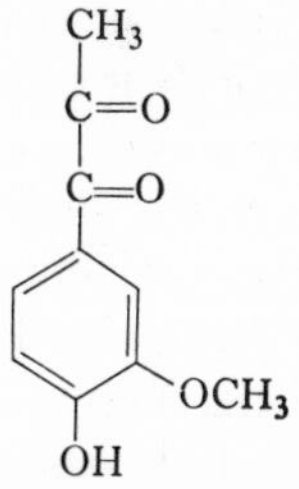

Lignin does not occur alone in nature, but rather it coexists with the cellulosic fraction of the cell wall components. Although it is intimately associated with these polysaccharides, this does not imply that lignin necessarily forms a chemical compound in the plant with the polysaccharides. It may merely exist in intimate physical association with them (9, p. 675; 10, p. 630). Indeed, much of the difficulty involved in studying the chemistry of lignin is attributable to the fact that it is so difficult to separate the lignin from the polysaccharides without applying reagents which are chemically so reactive that they inadvertently alter, to a greater or lesser degree, the structure of the lignin itself. The question of the exact nature of the lignin–polysaccharide complex is still undecided.

II. THE ISOLATION OF LIGNIN

Thus, much of the difficulty encountered by lignin chemists in their study of the structure of this material can be traced to the fact that, for a long time, no method was known by which lignin could be isolated as it exists *in situ*, i.e., in its natural form in the plant. That is to say, no matter what method of isolation was applied, the preparation obtained was not identical with the lignin as it had existed in the plant; the method of isolation itself had changed the chemical structure and properties of the lignin.

The various procedures described in the literature (9, p. 49; 10, p. 62) for the isolation of lignin may be divided into two classes: (a) those that depend on the removal by hydrolysis of the cellulose and other polysaccharide constituents of the wood by chemical treatment, leaving the lignin as an insoluble residue, and (b) those that depend on the separation of the lignin from the cellulose and other polysaccharides with which it exists by reagents which selectively dissolve the lignin.

However, as already mentioned, in addition to lignin and polysaccharides, all woods also contain greater or lesser amounts of certain "extraneous components," such as waxes, resins, organic acids, and pigments. To avoid the possibility of their contamination of the isolated lignin, these substances must first be removed by extraction with suitable solvents. The most advantageous solvents for such extraction are ether, benzene, ethyl alcohol, or mixtures of these solvents. Among the most frequently employed solvents for rendering the lignified plant material free of the extraneous materials is a mixture of (1 part) ethyl alcohol and (2 parts) benzene. This extraction is usually followed by washing the wood first with cold and then hot water. This preextracted wood may then be used as the source of lignin (36).

For the isolation of lignin by methods of the first of the above classes, the following reagents or procedures have been employed to hydrolyze or solubilize the cellulose: sulfuric acid (56), fuming hydrochloric acid (26), hydrofluoric acid (66), cuprammonium hydroxide (18a), periodic acid (65), and hydrogenolysis (40).

Many investigators have employed strong alkali for isolating lignin from woody tissues. Essentially, the method consists of dissolving the lignin by treatment with a basic solution and then acidifying the alkaline extract to precipitate the lignin. The ease with which the lignin can be obtained by this method depends somewhat on the nature of the lignified plant material. In the case of annual plants, the lignin may be obtained by treatment with alcoholic or aqueous sodium hydroxide solutions in the cold, whereas in the case of woods, a more drastic treatment, involving autoclaving under higher temperatures and pressures, is usually necessary (36).

III. BRAUNS' NATIVE LIGNIN

The methods for the isolation of lignin referred to above are similar to one another at least to the extent that an acid or a base, used per se or as a catalyst, is employed. Although the lignin is thereby obtained almost quantitatively, the reagent undoubtedly has some effect in altering the structure of the lignin polymer. Thus, it is generally accepted that even small amounts of acid promote the condensation of lignin (36).

In order to avoid this difficulty, F. E. Brauns, in 1939, employed 95% ethyl alcohol (*without* the traditional mineral acid catalyst) as the extraction solvent. His method involved extracting ground sprucewood meal with ethyl ether, and then with water at room temperature. This preextracted wood was then thoroughly reextracted with 95% ethyl alcohol, also at room temperature, until further alcoholic extracts no longer responded positively to the standard phloroglucinol-hydrochloric acid color test for lignin. Upon removal of the alcohol by distillation at a reduced pressure, a resinous material remained which was washed with water and ether. The resulting powder was dried, redissolved in dioxane, centrifuged, filtered, and precipitated by addition to a large volume of water. The precipitate was collected and dried *in vacuo*. This material was redissolved in dioxane and reprecipitated, this time into a large volume of ether. This purification procedure was repeated until a constant methoxyl value was obtained. The yield of lignin so obtained was 3% of the weight of the wood, or about 10% of the total lignin content, as determined by the procedure of Klason (7a).

Brauns found that his product had a methoxyl value of 14.8%, which was identical with the calculated value for spruce lignin as it exists *in situ* (7b). Moreover, this lignin preparation was found to be soluble in methanol, ethanol, dioxane, glacial acetic acid, acetone, pyridine, and dilute sodium hydroxide, and insoluble in diethyl ether, benzene, petroleum ether, and water. It reduced Fehling's solution and produced a red-purple color with Wiesner's reagent. Brauns named his preparation *spruce native lignin*. He later isolated other native lignins from western hemlock (8) and from aspenwood (13). Similar preparations have since been isolated from many species of softwoods (49), hardwoods (32), and from bagasse (60). Brauns' product has been criticized (18) as not really being lignin, and defended by its discoverer (10, p. 62).

IV. NORD'S ENZYMICALLY LIBERATED LIGNIN

Ever since the classic studies of Falck (17), it has been generally agreed that there are two different types of wood decay brought about by fungi, namely, the "brown" and the "white" rots. In the former, preferential attack is made on the carbohydrate components of the wood; the lignin remains essentially unaffected, and the decaying residue turns brown in color. In the

TABLE I

THE EFFECT OF THE ACTION OF BROWN-ROT ORGANISMS
ON THE CHEMICAL COMPOSITION OF WOOD (49)

Wood species	Organism	Decay period (months)	Cellulose (%)	Lignin (%)
White fir	*Lentinus*	0	60.3	24.9
	lepideus	4	51.2	30.6
		5	47.3	33.8
		6	46.5	36.4
		7	45.8	39.3
White Scots	*Lentinus*	0	61.7	26.8
pine	*lepideus*	4	49.5	31.8
		5	46.1	37.4
		6	46.0	40.0
		7	39.9	44.9
White Scots	*Poria*	0	45.5	33.9
pine	*vaillantii*	3	45.4	34.9
		5	33.4	42.1
		8	25.6	46.5
		11	17.6	51.1
		15	15.2	52.5
White Scots	*Lenzites*	0	45.5	33.9
pine	*sepiaria*	3	39.9	37.9
		6	30.1	41.0
		9	19.5	45.6
		13	18.5	50.1

latter, lignin seems to be the main substrate of the fungus, and, in the residue, there are patches of a white substance which has been considered to be a relatively pure form of cellulose. Accordingly, if an attempt were to be made to obtain lignin after the action of fungi on wood, organisms of the first group, i.e., the "brown" rots, must be employed, since these possess a cellulose-hydrolyzing enzyme system, but leave the lignin essentially unaltered.

Consequently, Schubert and Nord (50) inoculated ground, sterilized samples of certain species of softwoods with representative members of this class of wood-destroying fungi, and the effects of the decay of the wood by the microorganisms were progressively followed by chemical analyses of the resulting partially decayed wood. Typical results of the periodic analyses of the decaying wood are presented in Table I. From the analyses, it is evident that the net effect of the action of these brown-rot organisms on wood was indeed a depletion in the cellulose content of the wood, concomitant with an increase in the relative content of lignin. This lignin-enriched, decayed wood, therefore, offered a potential source for the isolation of a chemically unaltered lignin.

A primary objective of these investigations was the isolation and characterization of a lignin preparation unexposed to any chemical treatment (as is the case in the usual methods) by taking advantage of the cellulose-hydrolyzing activities of the brown-rot fungi. In practice, this product was obtained by extraction of the decayed wood with ethyl alcohol at room temperature, i.e., by the method of Brauns. However, a small amount of native lignin exists in sound wood and this can be obtained from the wood by the same extraction procedure. Accordingly, native lignin was expected to be present in an alcoholic extract of decayed wood, together with any lignin which might have been "freed", i.e., made more accessible to extraction by ethyl alcohol, as a result of the enzymic activity.

Indeed, a marked similarity was observed between the native lignin and the lignin obtained after the enzymic action of the wood-rotting fungi on white Scots pinewood. A comparison of the two preparations is presented in Table II. Thus, it was possible to obtain from the fungally decayed pinewood, by a mild extraction procedure, a lignin preparation that was very similar to native lignin. This lignin was isolated in yields superior to those obtainable from sound, uninfected wood, as seen in Table III. For example, after 15 months' decay of white Scots pinewood by the typical brown-rot fungus *Poria vaillantii*, up to 22.7% of the total lignin could be isolated as so-called "enzymically liberated lignin." Thus, it was proposed

TABLE II

A COMPARISON OF THE NATIVE AND ENZYMICALLY LIBERATED
LIGNINS OF WHITE SCOTS PINEWOOD (49)

Properties	Native lignin	Enzymically liberated lignin
C (%):	64.0	64.2
H (%):	6.3	6.0
OCH₃ (%):	14.5	14.4
OCH₃ of acetate (%):	10.1	10.2
OCH₃ of phenylhydrazone (%):	13.3	13.4
Soluble in:	EtOH, MeOH, dioxane, pyridine, 4% NaOH, glacial HOAc	Same
Insoluble in:	Water, ether, benzene, petroleum ether	Same
Reducing ability:	Reduces Fehling's solution	Same
Color reaction with		
Phloroglucinol:	Red-violet	Same
Aniline:	Yellow	
p-Phenylenediamine:	Yellow	Same
Diphenylamine:	Yellow	

that this lignin, isolated from fungally decayed wood, was "liberated" from
its association with the polysaccharide fraction of the wood as a direct result
of the enzymic activity of the fungus causing the rot (49).

TABLE III

THE YIELDS OF LIGNIN ISOLATED FROM SOUND AND
FROM DECAYED WHITE SCOTS PINEWOOD (49)

Period of decay (months)	Organism	Percentage of total
0	—	3.2
13	Lenzites sepiaria	18.3
15	Poria vaillantii	22.7

A comparison of this material with the small amount of native lignin isolable from sound wood indicated a close similarity between the two preparations (Table II). This finding therefore suggested that native lignin might indeed be identical with the total lignin content of the wood (49).

In another series of experiments, by first removing the native lignin from the wood by an exhaustive extraction until a negative phloroglucinol-hydrochloric acid reaction was repeatedly obtained on the final alcoholic extracts, and then subjecting this native lignin-free wood to fungal decay, it was possible to obtain lignin preparations that contained only that lignin which was "liberated" by the enzymic activity of the fungus causing the rot. Thus, it could be stated that such lignins resulted *exclusively* from the action of the fungi on the carbohydrate fraction of the wood. Such experiments were performed with the softwood white Scots pine (50), with the hardwoods, oak, birch, and maple (33), and with bagasse, the supporting fiber of the sugar cane plant (61). Some of the properties of the native and enzymically liberated lignins of these species are compared in Tables IV and V. The data shown in Tables IV and V reveal certain divergencies among the lignins of the various species, but they also reveal the apparent identity of the native lignins with the enzymically liberated lignins of the individual species studied (48).

TABLE IV

A COMPARISON OF THE NATIVE (N.L.) AND ENZYMICALLY
LIBERATED (E.L.) HARDWOOD LIGNINS (32, 33)

Properties	Oak		Birch		Maple	
	N.L.	E.L.	N.L.	E.L.	N.L.	E.L.
C (%):	58.6	58.4	61.4	61.6	61.0	61.3
H (%):	5.3	5.2	5.5	5.6	5.6	5.5
OCH$_3$ (%):	14.8	14.6	14.9	14.8	17.4	17.8
OCH$_3$ of phenylhydrazone (%):	13.7	13.6	13.1	13.4	15.5	15.8
OCH$_3$ of acetate (%):	10.2	10.3	10.2	10.4	12.8	12.9
OCH$_3$ of CH$_2$N$_2$-methylated lignin (%):	25.0	24.8	26.0	25.8	23.6	23.9
Oxidation of lignin Vanillin (%):	21.3	20.9	18.6	19.1	17.2	16.9
Syringaldehyde (%):	0	0	0	0	4.5	4.2

TABLE V

A COMPARISON OF NATIVE (N.L.) AND ENZYMICALLY
LIBERATED (E.L.) LIGNINS FROM BAGASSE (61)

Properties	N.L.	E.L.
C (%):	61.5	61.6
H (%):	5.7	5.9
OCH$_3$ (%):	15.3	15.4
OCH$_3$ of acetate (%):	13.3	13.1
OCH$_3$ of phenylhydrazone (%):	14.1	14.0
Oxidation products		
Vanillin (%):	17.8	17.3
Syringaldehyde (%):	13.3	12.9
p-Hydroxybenzaldehyde (%):	9.8	10.2

The enzymically liberated lignins were found to be identical with the native lignins of the same species in regard to their elementary compositions, the compositions of their derivatives, solubilities, reducing ability, color reactions, and ultraviolet and infrared absorption spectra. Thus, it appeared that the lignin of each species might be a uniform chemical substance or group of related substances, of which the greater part, the "extra-native" lignin, was associated with the polysaccharide fraction in such a way that its extraction from sound wood by chemically inert solvents, such as ethyl alcohol (35), would be rendered impossible or at least very difficult.

V. BJØRKMAN'S MILLED-WOOD LIGNIN

The possibility of obtaining larger amounts of a chemically unchanged lignin from very finely divided wood had been considered for some time. Thus, Brauns and Seiler (11) homogenized sprucewood by grinding the wood powder as an aqueous suspension in a Valley beater and as a result, they were able to obtain extremely fine particles. However, these particles did not yield any lignin in addition to the already extracted native lignin; this was also found to be true for other wood species (53).

Subsequently, A. Bjørkman found that by grinding sprucewood meal (which was dispersed in a nonswelling solvent such as toluene) in a vibrational ball mill, it was possible to extract with aqueous dioxane 50% or more of the lignin of the wood (4). This product is referred to as "milled-wood lignin" and is a faintly cream-colored, ash-free powder with an average molecular weight of about 11,000.

The data that have been obtained on analysis of milled-wood lignin indicate that the original structure of the material may be better preserved in it than in many other lignin preparations. Thus it has been reported that milled-wood lignin is a phenylpropane-derived polymer containing free phenolic hydroxyl groups, p-hydroxybenzyl alcohol moieties, aromatic ring-conjugated carbonyl groups, and possibly ethylenic double bonds (5). The sharpness of the absorption bands of its infrared spectra is alleged to be an indication of the homogeneity of this lignin. Milled-wood lignins have also been prepared from other species of softwoods (37) as well as from hardwoods (15).

VI. CRITERIA FOR LIGNIN HOMOGENEITY

It is obviously possible that any individual lignin preparation that has been isolated from a given species of wood by a certain method is not entirely homogeneous, but is in fact structurally a mixture of related molecules; this is true of the native lignins of oak and birch (33). Thus the methoxyl content of the native lignin fractions of these wood species was found to be about 15%, whereas the Klason lignins of the same woods were found to have 20% methoxyl.

As a result, lignins should perhaps be regarded as mixtures of molecules, all possessing similar chemical structures, but with the possibility of certain structural differences. It is to such mixtures that may differ both in structure and in molecular weight that the notion of *Gruppenstoffe* was first applied (59). Thus they are distinct from isomeric compounds that have the same molecular weight but differ in structure, or from homologs that have different molecular weights but essentially the same structure.

The characterization of such substances by the classic methods of organic chemistry presents many difficulties. Hence, in the field of lignin chemistry, one had to rely mainly on elementary composition, methoxyl content, and the properties of derivatives and degradation products. To overcome these limitations, physicochemical methods are now employed, and spectroscopic analyses, including ultraviolet and infrared absorption spectra, have been applied with success (49, 50).

Because of the success in the application of electrophoretic methods to the characterization and fractionation of proteins, an investigation of the possible use of such methods in studying lignin was carried out (54). Accordingly, the electrophoretic patterns of various native and enzymically liberated lignins were obtained. Thus it was observed that certain lignin preparations appeared perfectly homogeneous, while other samples presented a smaller boundary of lower mobility, or "trailing boundary," next to the main boundary.

As an illustration of electrophoretically homogeneous boundaries, the electrophoretic patterns obtained for the lignin of cork are presented in Fig. 1. Homogeneous patterns were also obtained with the native maple lignin.

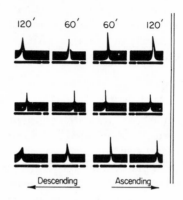

FIG. 1. Electrophoretic patterns of fractionated cork lignins and of Indulin (pH 10.7, 0.2 M NaCl, 0.05 M sodium glycinate-HCl).

On the other hand, native white Scots pine lignin, although reprecipitated several times, presented a well-marked trailing boundary, as shown in Fig. 2. It is noteworthy that the same trailing boundary was also observed in samples of enzymically liberated white Scots pine lignin. The situation was similar with native and enzymically liberated oak lignin. Native bagasse lignin appeared to be somewhat inhomogeneous and presented a detectable

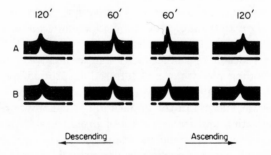

FIG. 2. Electrophoretic patterns of lignins. A, White Scots pine enzymically liberated lignin. B, Mixed samples of white Scots pine native and enzymically liberated lignin (pH 10.7, 0.2 M NaCl, 0.05 M sodium glycinate-HCl).

second boundary. Enzymically liberated bagasse lignin behaved similarly. The results of the electrophoretic analyses are summarized in Table VI. Mobilities of the main boundaries only are presented. It is evident that, under similar conditions of pH and buffer composition, all of the lignin samples present comparable mobilities (54).

TABLE VI

ELECTROPHORETIC MOBILITIES OF VARIOUS NATIVE
LIGNINS IN GLYCINE-NaCl BUFFER, pH 10.7 (54)

Lignin samples	Mobility (cm.^{2}/volt/sec. × 10^5)	
	Ascending	Descending
White Scots pine	−10.2	−10.1
Bagasse	−9.8	−9.1
Maple	−9.3	−8.3
Oak	−8.7	−8.2
Cork	−9.5	−8.9

In view of the complexity of the lignins and the possibility of variations in their structures, it was interesting to find that the majority of the lignins studied did give patterns characteristic of electrophoretically homogeneous materials; the most significant exception was white Scots pine lignin. It was found that this lignin, whether native or enzymically liberated, presented the same pattern, and even when the samples were mixed, the electrophoretic pattern did not change. Therefore, it was assumed that this lignin is a mixture of electrophoretically distinct components whose rate of extraction is the same before and after fungal attack (54).

The heterogeneity of these native and enzymically liberated lignins was then investigated by means of paper chromatography (51), whereby "mobile" and "immobile" components were detected. Consequently, lignins must be regarded as mixtures of components, all of which possess similar structures, but with the possibility of certain minor chemical differences also present, i.e., substances of the type for which Staudinger introduced the term *Gruppenstoffe* (59).

VII. CHEMICAL CRITERIA FOR LIGNIN PREPARATIONS

Since lignin does not appear to have one uniform chemical identity, a chemical definition for it is difficult to formulate. However, Kratzl and Billek (28) have listed eight chemical features that constitute "criteria" by which a lignin sample may be characterized. These features will be described and will be illustrated with data for specific lignin preparations obtained from various wood sources.

A. Solubility in Various Solvents

In Table VII, the solubilities in various solvents of the native and enzymically liberated lignins of white Scots pinewood are presented, as well as of the

TABLE VII

SOLUBILITIES OF LIGNINS FROM WHITE SCOTS PINEWOOD[a] (47)

		Lignins			
Solvents	Native	Enzymic-ally liberated	Isolated with H₂SO₄	Isolated with fuming HCl	Isolated with alkali
Methanol	S	S	I	I	I
Ethanol	S	S	I	I	I
Dioxane	S	S	I	I	S
Pyridine	S	S	I	I	S
4% NaOH	S	S	I	I	S
Acetic acid	S	S	I	I	I
Water	I	I	I	I	I
Ether	I	I	I	I	I
Benzene	I	I	I	I	I
Petroleum ether	I	I	I	I	I

[a] S = Soluble;   I = Insoluble.

lignins isolated from this wood by treatment with 72% sulfuric acid, fuming hydrochloric acid, and 10% sodium hydroxide. None of these lignins is soluble in water or nonpolar organic solvents. In general, the lignins that are isolated by extraction with organic solvents (i.e., native and enzymically liberated lignins) are soluble in more polar organic solvents, while those that are obtained by chemical treatment are insoluble in these solvents, except for "alkali lignin" which is soluble in dioxane, pyridine, and dilute base (47).

B. Ultraviolet and Infrared Absorption Spectra

With the exception of aspen native lignin, bagasse native and enzymically liberated lignins, and "kiri" native lignin, all the native lignins investigated and their derivatives have one maximum at about 280 mμ. Certain of these maxima are tabulated in Table VIII. The absorption curves of some typical native and enzymically liberated lignins are presented in Fig. 3 (36).

TABLE VIII

ABSORPTION MAXIMA AND EXTINCTION COEFFICIENTS
OF VARIOUS LIGNINS (36)

Type of lignin	Solvent	Maximum (mμ)	$E_{1\ cm.}^{1\%}$
Spruce, native	Ethanol	280	—
Hemlock, native	Ethanol	280	—
White Scots pine, native and enzymic	Dioxane	282	204
Oak, native and enzymic	Dioxane	282	195
Birch, native and enzymic	Dioxane	280	120
Maple, native and enzymic	Dioxane	280	150
Aspen, native	Dioxane	None	—
Kiri, native	Dioxane	280	137

The maximum at 280 mμ persists in spite of such alterations in the material as are caused by methylation, acetylation, and treatment with sodium hydroxide. However, the absorption of the phenylhydrazones of the native lignins of spruce (19) and of white Scots pine (63) are atypical in that a second maximum occurs at 352 mμ. When the phenylhydrazone is formed, however, the keto groups are "fixed" in the carbonyl structure, and there appear to be enough of these groups present to produce a definite absorption band.

However, Nord and deStevens (36) felt that this interpretation of the effect of an enolizable carbonyl group was inadequate. They found (63) that, although the phenylhydrazone derivative of white Scots pine native lignin gave an absorption maximum at 352 mμ, nevertheless, the carbonyl group of the native lignin was not enolized. Furthermore, they noted that the phenylhydrazone of the diazomethane-methylated white Scots pine native lignin did give some absorption at 352 mμ while the methylated lignin did not. Thus, they felt that it was more correct to say that absorption at 352 mμ

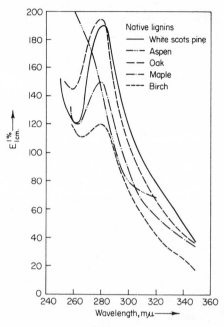

FIG. 3. Absorption curves of some typical native lignins (solvent, 90% dioxane).

is not governed by either the presence or absence of an enolized carbonyl group, but by the relative strength of the carbonyl absorption band at that wavelength. The phenylhydrazone of the native lignin seemed to intensify this absorption band.

Ultraviolet spectroscopic studies on bagasse native lignin demonstrated that this lignin seemed to have a constitution somewhat different from other lignins (60). Thus, it gave an absorption plateau between 282 and 295 mμ and an absorption peak at 315 mμ. It has been demonstrated that absorption above 300 mμ is due to the presence of a chromophoric group conjugated with an oxygen-substituted benzene ring (36). To determine the nature of this group (i.e., carbonyl or ethylenic double bond), bagasse native lignin was hydrogenated in the presence of Raney nickel at temperatures from 25° to 50°C. and relatively low hydrogen pressures (2–10 atm.). Under these conditions, ethylenic double bonds of an aliphatic chain are readily reduced (36). But here, the hydrogen uptake was nil, and the nature of the ultraviolet absorption spectrum did not change. Moreover, there was no change in the methoxyl content after the attempted hydrogenation, and the preparation still responded to the phloroglucinol-hydrochloric acid color test. The authors thus concluded that the chromophore was a carbonyl group (36).

The infrared absorption spectrum of spruce native lignin was first determined by Jones (25). He found it to be generally similar to those of lignins isolated by strong chemical reagents. Since this initial study, the infrared spectra of the native and enzymically liberated lignins from white Scots pine (50), oak (32), birch (32), maple (32), bagasse (60), and kiri wood (62) have been determined. In all cases, they were found to be quite similar to the spectrum of spruce native lignin. Some of these spectra are presented in Fig. 4.

The band at 3400 cm.$^{-1}$ is due to the presence in lignin of bonded hydroxyl groups. However, the intensity of the absorption band in this region is less pronounced in the case of bagasse native lignin than in the other lignin preparations. This supports the chemical data which indicated that bagasse native lignin contains fewer hydroxyl groups than the other lignins. Absorption from 1700 to 1660 cm.$^{-1}$ is characteristic for the presence of an aldehyde or ketone carbonyl group. This band is very weak in the case of maple lignin. Phenyl ring skeletal vibrations with possible *para* substitution are indicated at 1660 and 1510 cm.$^{-1}$. The 1430-cm.$^{-1}$ band establishes the presence of aliphatic structures. A difference in relative intensity at 1325 cm.$^{-1}$ should also be pointed out. Such variations are more evident at 1265, 1255, and 1220 cm.$^{-1}$, which are in the spectral region of aromatic or aliphatic unsaturated C—O groups, and also at 1140, 1163, 1155, and 1125 cm.$^{-1}$. Since bands in this region arise from vibrations of the molecule as a whole, interpretation cannot be specific. The bands at 2945, 1460, 1380, and 725 cm.$^{-1}$ are of no interpretative value (36).

Bagasse native lignin and kiri native lignin have sharp bands at 835 cm.$^{-1}$, while maple lignin has pronounced bands at 890 and 870 cm.$^{-1}$. Using a film technique (evaporation of a dioxane solution of spruce native lignin), Jones (25) reported bands in the same region, even after prolonged drying. (Dioxane is known to absorb in this part of the spectrum.) But since mineral

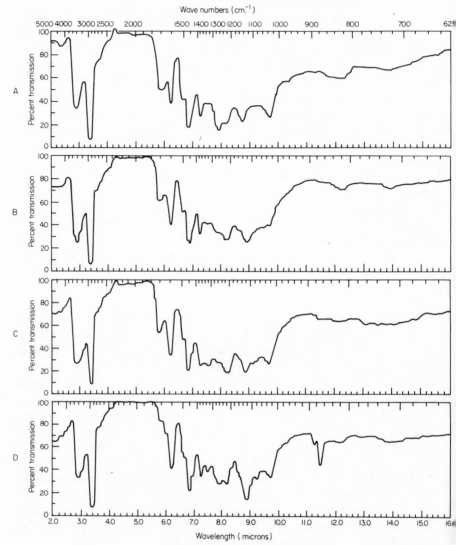

FIG. 4. Infrared absorption spectra of some native lignins: A, white Scots pine; B, birch; C, oak; D, maple.

oil mulls were used with the other lignins referred to here, it may be stated that the bands were characteristic of the lignins themselves. These bands are suggestive of strong absorption by trisubstituted phenyl groups.

Thus, a comparative evaluation of the infrared absorption spectra of these lignins revealed that they were generally similar, but noticeable differences were observed in the relative intensities of the bands; this suggested some deviation in the arrangement of the groups comprising the total lignin structure. Moreover, it was also found that the spectra of the enzymically liberated lignins from each individual species of wood appeared identical with the corresponding native lignin fraction from that same wood species (36).

C. Determination of Functional Groups

The functional groups of lignin include methoxyl, phenolic hydroxyl, primary and secondary alcoholic hydroxyl, benzyl alcohol groups, ether groups (aryl ethers and coumaran structures), carbonyl, carboxyl, and conjugated carbonyls and other double-bonded groups.

The elementary contents and a (partial) functional group analysis of the native and milled-wood lignins from pinewood and sprucewood are presented in Table IX (22), as illustrations of the compositions of typical isolated lignin preparations.

D. Oxidative Degradation

1. With Nitrobenzene and Alkali

The application of oxidation reactions to the elucidation of the structure of natural products has had wide utility, and the usefulness of this practice has been firmly established. Numerous oxidizing agents have been employed in the degradation of lignin but the mildest is probably nitrobenzene with alkali. The degradation products usually obtained by this method are vanillin (III) from softwood lignins, vanillin and syringaldehyde (IV) from hardwood lignins and p-hydroxybenzaldehyde (V) from perennial plant lignins.

Paper partition chromatography has been applied to the separation, identification and estimation of these aldehydes (64). The quantitative results of such analyses performed on several native and enzymically liberated lignins are presented in Table X. The lignins from the annual plant, sugar cane, and from the softwood, white Scots pine, gave rise to vanillin as well as to significant amounts of p-hydroxybenzaldehyde, whereas the lignins from birch yielded only traces of this substance. Investigation of native and enzymically liberated lignins from oak and birch revealed that these relatively low methoxyl-containing hardwood lignins yielded only small amounts of syringaldehyde (63).

TABLE IX

ANALYSES OF PINE AND SPRUCE LIGNINS (22)

| Lignin | C (%) | H (%) | N (%) | MeO (%) | Vanillin (%) | Hydroxyl | | | Car-bonyl | Car-boxyl | $\epsilon_{max.}/\epsilon_{min.}$ |
| | | | | | | Total | Phenolic | | | | |
							(a)	(b)			
Pine, N.L.[c]	63.85	6.22	0.0	14.5	22.6	—	0.27	0.32	0.24	0.022	1.31
Spruce, N.L.[c]	63.57	6.32	0.0	14.8	21.2	1.15	0.30	0.32	0.23	0.021	1.39
Pine, M.W.L.[d]	63.70	6.29	0.0	15.5	20.0	—	0.22	0.24	0.20	0.024	1.21
Spruce, M.W.L.[d]	63.15	6.21	0.0	15.9	19.9	1.20	0.24	0.25	—	0.030	1.28

[a] Determined by $\Delta\epsilon$ method.
[b] Determined by diazomethane method.
[c] N.L., native lignin.
[d] M.W.L., milled-wood lignin.

TABLE X

YIELDS OF OXIDATION PRODUCTS OF LIGNINS (63)

Lignins	p-Hydroxy-benzaldehyde (%)	Vanillin (%)	Syringal-dehyde (%)
White Scots pine, native	3.5	19.5	—
White Scots pine, enzymic	2.4	18.7	—
Oak, native	—	21.3	4.0
Oak, enzymic	—	20.9	3.6
Birch, native	>0.5	18.6	1.7
Birch, enzymic	>0.5	19.1	2.0
Maple, native	—	17.2	4.5
Maple, enzymic	—	16.9	4.2
Bagasse, native	9.8	17.8	13.3
Bagasse, enzymic	10.2	17.3	12.9

2. WITH PERMANGANATE AFTER METHYLATION OF THE LIGNIN

Employing potassium permanganate as the oxidizing reagent, Richtzenhain (44) oxidized the methylated forms of sprucewood, Willstätter lignin, lithium lignosulfonate, soluble ethanol lignin, the wood residue from the ethanol lignin preparation, a Willstätter lignin treated with 70% potassium hydroxide at 170°C., and a sulfate lignin. The oxidation products obtained included veratric acid (VIII), dehydrodiveratric acid (IX), isohemipinic acid (X), and metahemipinic acid (XI).

The observation that wood itself, i.e., "protolignin," gave small amounts of isohemipinic acid seemed to indicate that the structure (XII) was an integral part of the lignin molecule. The fact that metahemipinic acid was obtained only from acid lignins seemed to indicate that some carbon-to-carbon condensation at the 6 position of the benzene ring had occurred as a result of the action of the acid. Since no metahemipinic acid was found in the oxidation products of "protolignin," and since thiolignin, which had not been exposed to acid treatment, also failed to yield this product, it was questionable whether or not the carbon structure of metahemipinic acid occurred as such in protolignin, or was formed in a secondary reaction by the action of the strong mineral acid. Richtzenhain (45) postulated that the formation of metahemipinic acid was a result of the presence in the lignin molecule of a "lignan"-type structure (XIII) which, under the influence of

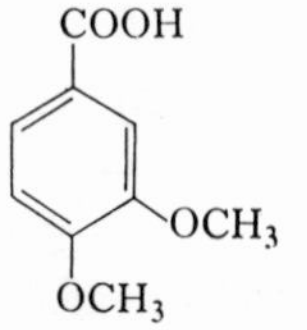

COOH

OCH$_3$

OCH$_3$

(VIII)

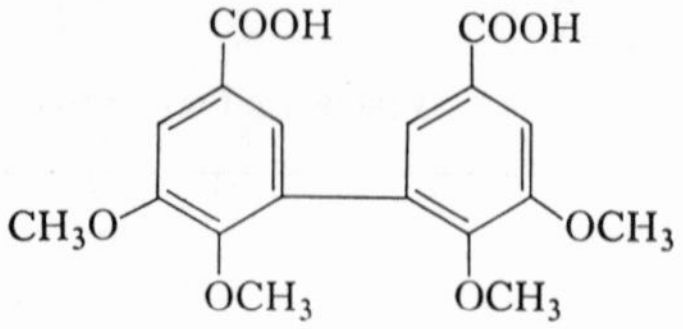

COOH COOH

CH$_3$O   OCH$_3$

OCH$_3$ OCH$_3$

(IX)

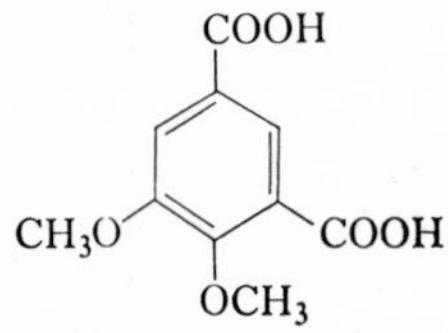

COOH

CH$_3$O   COOH

OCH$_3$

(X)

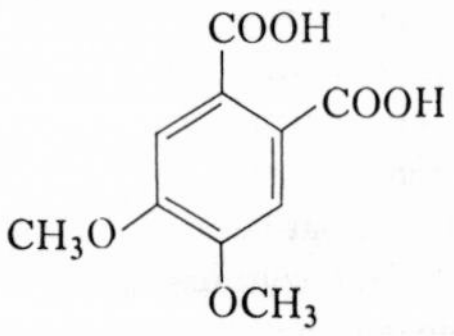

COOH

COOH

CH$_3$O

OCH$_3$

(XI)

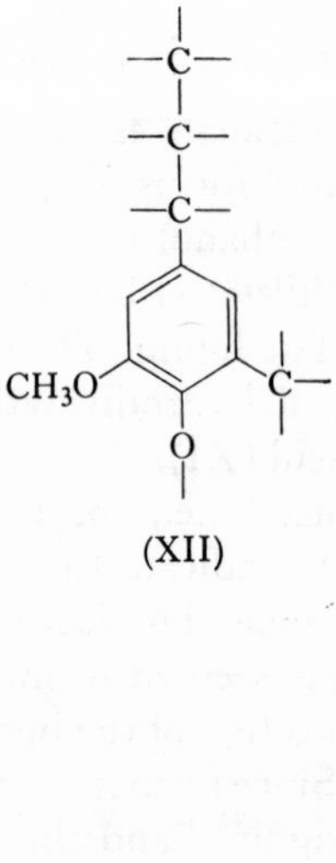

—C—

—C—

—C—

CH$_3$O   —C—

O—

(XII)

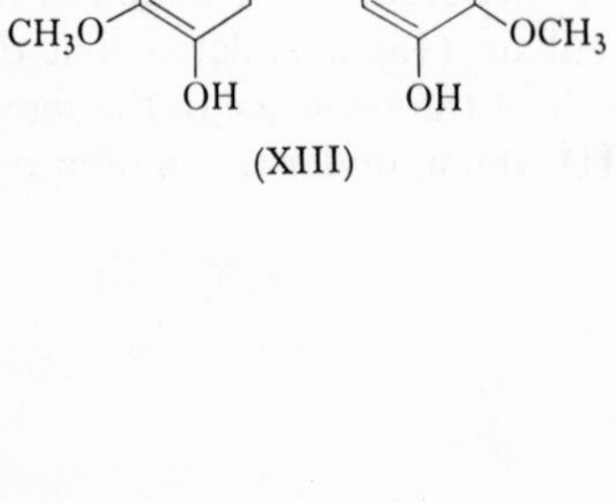

—C— —C—

—C———C—

—C— —C—

CH$_3$O   OCH$_3$

OH OH

(XIII)

—C— —C—

—C———C—

—C— —C—

CH$_3$O   OCH$_3$

OH OH

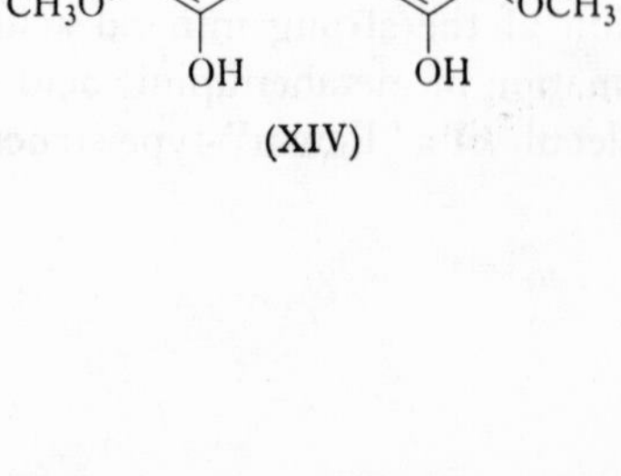

(XIV)

the acid used in the isolation of the lignin, was converted into an "isolignan"-type structure (XIV). Mikawa (34) also obtained veratric and isohemipinic acids upon oxidation of methylated thiolignin with potassium permanganate.

3. WITH PERMANGANATE DIRECTLY

A study of the formation of benzenepolycarboxylic acids from lignin on its direct oxidation with potassium permanganate was made by Read and Purves (43). They oxidized preextracted sprucewood, the periodate, hydrochloric acid, and Klason lignins from the same wood, and a commercial alkali lignin from aspenwood with potassium permanganate in 1% aqueous potassium hydroxide at room temperature and at 75°–80° C. The benzenepolycarboxylic acids that formed were separated by a specially developed method, and were found to include benzene-1,2,4,5-tetracarboxylic acid (XV), benzenepentacarboxylic acid (XVI), and benzene-hexacarboxylic acid (XVII). The quantitative results seemed to indicate that

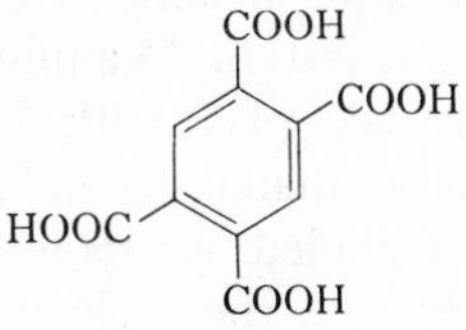

(XV)

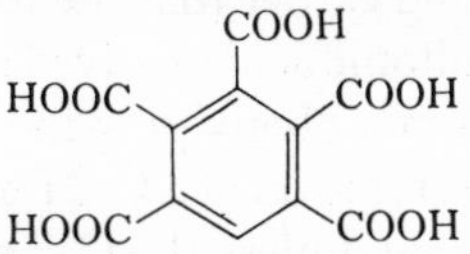

(XVI)

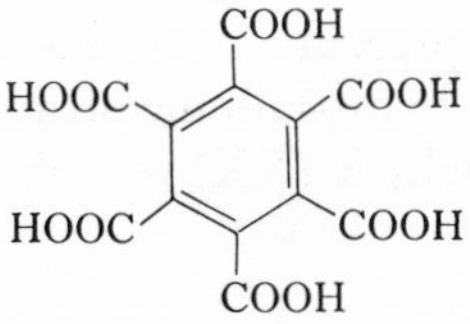

(XVII)

aromatic lignin building stones underwent an increasing degree of nuclear condensation as the conditions of isolation from the wood became more drastic; this was true not only under acid conditions, but also under alkaline. The formation of these polycarboxylic acids was explained as resulting from a conidendrin-like (XVIII) structure in which the two aromatic rings are degraded, with the cycloaliphatic ring, after dehydrogenation, yielding the benzenepolycarboxylic acids on its oxidation.

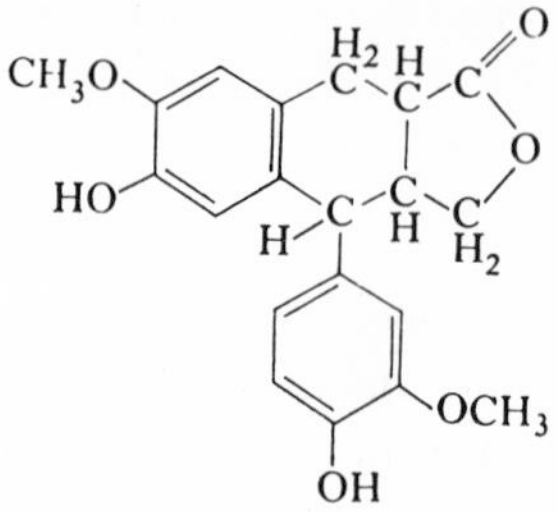

(XVIII)

E. Alkaline Hydrolysis

To gain insight into the nature of the propyl side chains of the lignin building stones, and the formation of vanillin, acetovanillone, and other carbonyl compounds, Kratzl and Hofbauer (29) carried out an extensive study of the alkaline hydrolysis of lignosulfonic acids. Buckland *et al.* (14) had suggested earlier that the formation of vanillin on the alkaline hydrolysis of lignosulfonic acid involved a "reversed aldol condensation." Vanillin formation from a phenylpropane building stone must take place with the simultaneous removal of a C-2 fragment which might be ethanol, acetaldehyde, or acetic acid, unless the C-2 moeity was further degraded. Kratzl found that, on the alkaline hydrolysis of lignosulfonic acid with the exclusion of air, acetaldehyde was formed; this result has also been obtained by Adler and his co-workers (2). The cleavage occurred very slowly. Therefore, Kratzl suggested that the lignin building stone from which the vanillin and acetaldehyde had originated was present in the lignin molecule in the form of a "masked coniferaldehyde hydrosulfonic structure" (XIX) from which α-hydroxydihydroconiferaldehyde (XX) was liberated and, by the "reversed

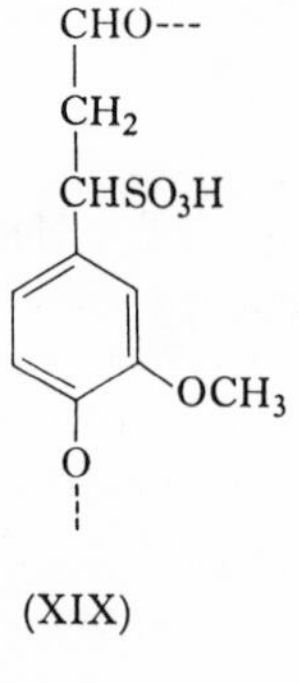

(XIX) 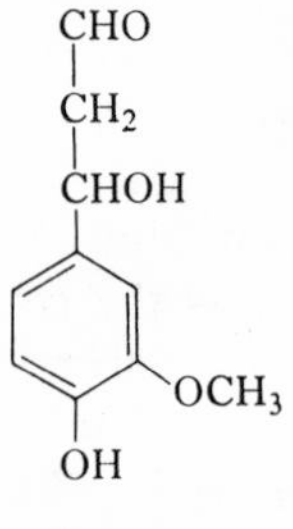   (XX)

aldol reaction," this was converted into vanillin and acetaldehyde. Kratzl and Keller (30) also subjected other lignosulfonic acids, prepared from hydrochloric acid spruce, cuproxam beech, and soluble native spruce lignins, to a stepwise alkaline hydrolysis and again obtained vanillin and acetaldehyde.

F. Cleavage with Sodium in Liquid Ammonia

Since the building stones in the structure of lignin may be combined with each other by ether linkages, particularly by aryl—alkyl ether bonds, Shorygina and Kefeli (58) subjected certain lignin preparations to the effect of sodium in liquid ammonia, since ether linkages are known (57) to be cleaved by this reagent. For example, cuproxam spruce lignin was treated with sodium in liquid ammonia, and at the low temperature employed, the reaction occurred very slowly, and after nine such treatments was 88.7% of the lignin degraded to low molecular weight compounds. From the reaction products, Shorygina was able to isolate dihydroeugenol (XXI) and 1-guaiacylpropanol-2 (XXII) in good yields.

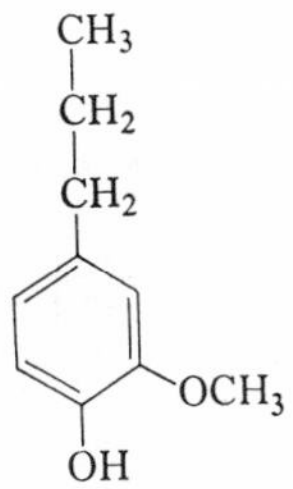

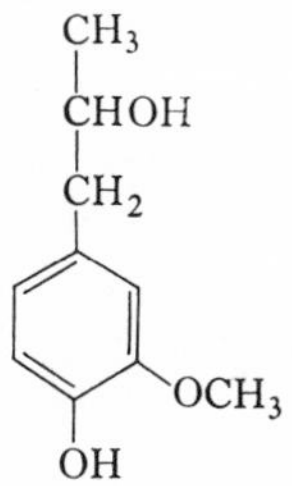

G. Ethanolysis

The prolonged heating of lignified materials with ethanol, or some other alcohol, plus hydrochloric acid, results in the "alcoholysis" or "ethanolysis" of the lignin. Brauns (9, p. 453) has noted that in the alcoholysis of lignin three reactions may take place simultaneously: (a) condensation of part of the lignin with the alcohol, resulting in a soluble "alcohol lignin," (b) condensation of part of the lignin with itself, resulting in a product of lesser reactivity, and (c) an "alcoholytic cleavage" of part of the lignin, resulting in its depolymerization to smaller lignin building units, which may be followed by the condensation of some of these units with the reacting alcohol.

In recent years, alcoholysis, and especially ethanolysis, has found use as an important analytical tool in helping to establish whether or not an isolated product is really a lignin. Thus, on treating "protolignin" with absolute alcohol in the presence of catalytic amounts of hydrochloric acid, Hibbert and his co-workers (21) obtained a series of monomeric phenylpropane derivatives. These compounds, which have come to be known as "Hibbert's ketones," are vanillin (III), α-ethoxypropioguaiacone (VI), vanilloyl acetyl (VII), and guaiacyl acetone (XXIII). Since these monomers seem to result from a specific reaction of certain building stones in the lignin structure, the reaction is now often used as a "test" for lignin preparations. Thus, Kratzl

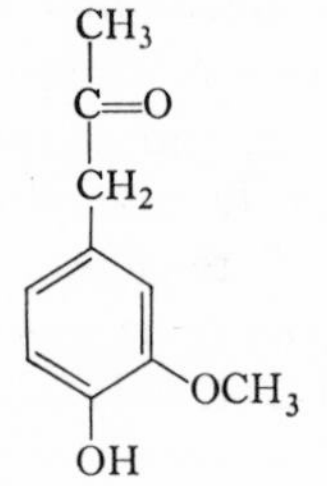

$$\begin{array}{c} CH_3 \\ | \\ C\!=\!O \\ | \\ CH_2 \end{array}$$

(XXIII)

and Klein (31) have applied the reaction to the characterization and identification of certain lignin preparations and have also developed a micromethod for its application subsequent to the administration of radioactively labeled compounds to living plants to determine their incorporation (or lack of same) into the lignin.

H. Hydrogenolysis

Of all the degradative methods applied to lignin to help gain insight into its structure, catalytic hydrogenation has been among the most useful, and it has provided strong evidence for the hypothesis that lignin consists of phenylpropane building stones. Thus, Harris et al. (20), Pepper and Hibbert (41), and others (9, p. 514; 10, p. 486) have isolated phenylpropane, or cyclohexylpropane derivatives, or both, as products of the high-pressure catalytic hydrogenation of lignin. The early workers subjected substances such as Meadol (1), methanol, and hydrochloric acid lignins (20), and even wood itself (6) to hydrogenation. Since the yields of identifiable reaction products were generally low, i.e., 20 to 40%, and since complicated mixtures were usually obtained, large amounts of starting materials were hydrogenated in order to permit the isolation and identification of the products.

For example, after the high-pressure copper chromite-catalyzed hydrogenation of aspen methanol lignin, Harris and co-workers (20) isolated the cyclohexylpropane derivatives, 4'-hydroxycyclohexylpropane (XXIV), 3',4'-dihydroxycyclohexylpropane (XXV), and 4'-hydroxycyclohexylpropanol-1 (XXVI). Hibbert and his co-workers (21) obtained the same products

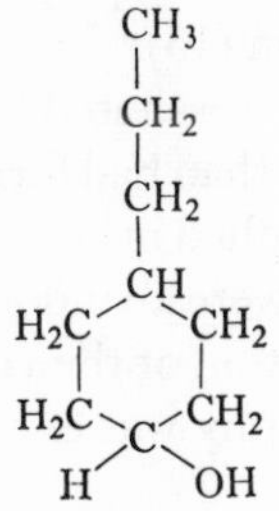

(XXIV) 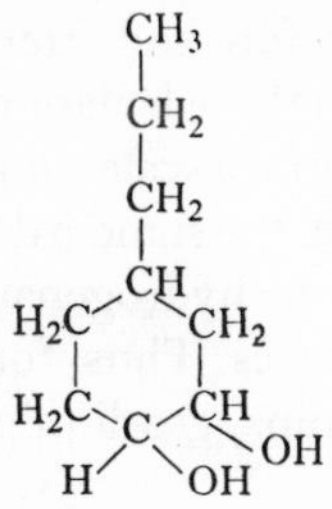   (XXV) 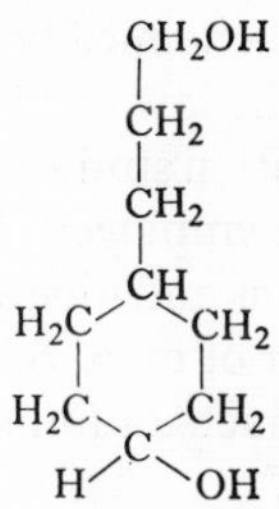   (XXVI)

after a similar treatment of maplewood and sprucewood and of maple ethanol lignin. However, when Raney nickel was used as the catalyst, and lower temperatures were employed, the reaction could be carried out so that the aromatic rings of the lignin remained unhydrogenated. Thus, by performing the reaction in this way, Brewer *et al.* (12) were able to isolate, from hydrogenated maple lignin, the aromatic compounds 3'-methoxy-4'-hydroxyphenylpropanol-1 (XXVII), 3',5'-dimethoxy-4'-hydroxyphenylpropane (XXVIII), and 3',5'-dimethoxy-4'-hydroxyphenylpropanol-1 (XXIX).

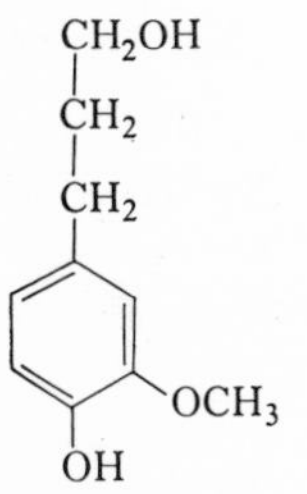

(XXVII) 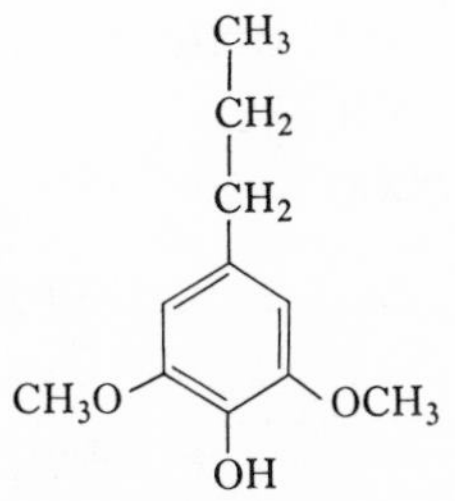   (XXVIII) 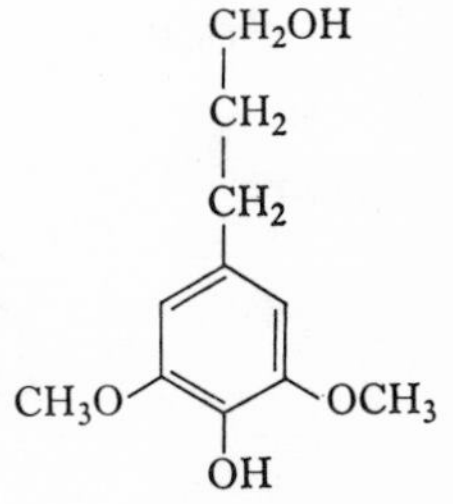   (XXIX)

However, until relatively recently, no studies of this type had been reported on lignin preparations that had been isolated by a method mild enough to preclude any chemical alteration before the hydrogenation. Furthermore, lignin preparations isolated under these mild conditions, such as native lignin and enzymically liberated lignin, are usually obtained in relatively low yields. A possible exception is milled-wood lignin, which has been isolated in yields of 60%. Accordingly, in a series of studies on the catalytic hydrogenation of certain softwood and hardwood lignins conducted in Nord's laboratory, native and milled-wood lignins were employed and were chemically compared before and after hydrogenation (15).

The low yields obtained in these isolations, however, necessitated hydrogenation studies of the lignins on a scale somewhat smaller than had formerly been employed. Furthermore, the standard techniques applied in earlier isolations and identifications of the hydrogenation products were abandoned in favor of more refined procedures. Thus, for the investigation of the volatile products of milled-wood lignins, gas-liquid chromatography was exploited (16).

Gas chromatography is used frequently in organic chemistry for the separation as well as the identification and quantitative determination of various compounds. The identification of products that are separated by this method is often realized by a comparison of their retention times with those of previously described compounds. Infrared and mass spectroscopy are frequently applied as supplementary aids to gas chromatography, but identification can sometimes be accomplished by the latter technique alone

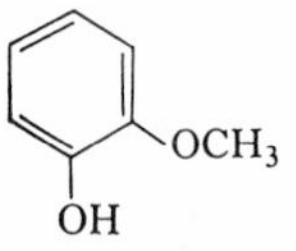

(XXX) 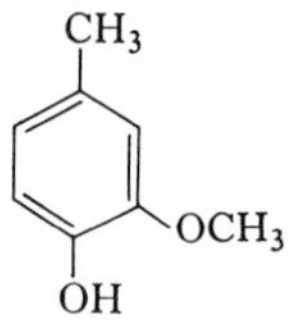   (XXXI) 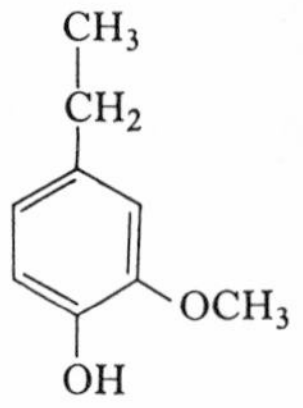   (XXXII)

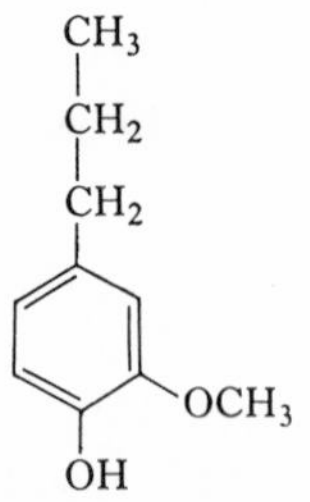

(XXXIII) 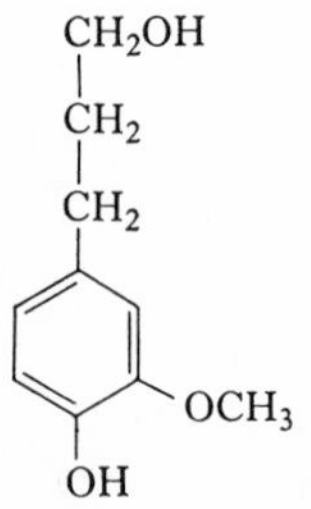   (XXXIV)

by a comparison of retention times on three different stationary phases, i.e., electron donor, electron acceptor, and nonpolar (37).

Such techniques were employed on the milled-wood lignins of two softwoods, namely, white pine and blue spruce (37). Subjection of the hydrogenation mixtures to gas-liquid chromatography revealed the presence of

TABLE XI

YIELDS AND MOLAR RATIOS OF SOFT-WOOD LIGNIN
HYDROGENATION PRODUCTS (37)

	Blue spruce			White pine		
Compounds	Distill-able (%)	Lignin (%)	Molar ratios[a]	Distill-able (%)	Lignin (%)	Molar ratios[a]
Guaiacol	1.0	0.3	1.0	1.0	0.3	1.0
4-Methylguaiacol	11.5	3.5	10.4	10.2	2.9	9.1
4-Ethylguaiacol	7.0	2.1	5.7	6.5	1.9	5.4
4-n-Propylguaiacol	19.9	5.9	14.9	18.3	5.3	13.7
Dihydroconiferyl alcohol	27.0	8.1	18.4	25.0	7.3	16.6
	66.4	19.9		61.0	17.7	

[a]Guaiacol assigned a value of unity.

guaiacol (XXX), 4-methylguaiacol (XXXI), 4-ethylguaiacol (XXXII), 4-n-propylguaiacol (XXXIII), and dihydroconiferyl alcohol (XXXIV), all of which were isolated, and identified by their retention times and infrared spectra. The yields and "molar ratios" of the products from the two softwood lignins are presented in Table XI. Similarly, subjection of the milled-wood lignins of two hardwoods, namely, birch and oak (16), also gave rise to (XXXI), (XXXII), (XXXIII), and (XXXIV), and also 4-methylsyringol (XXXV), 4-ethylsyringol (XXXVI), 4-n-propylsyringol (XXXVII), and dihydrosinapyl alcohol (XXXVIII). The yields and "molar ratios" of the products from the two hardwood lignins are presented in Table XII.

To help elucidate the nature of the reactions that resulted in the formation of these compounds during the catalytic hydrogenation of the milled-wood lignins, "model compounds" were synthesized, containing hydroxyl groups, carbonyl groups, ethylenic double bonds, and aryl-ether linkages. These

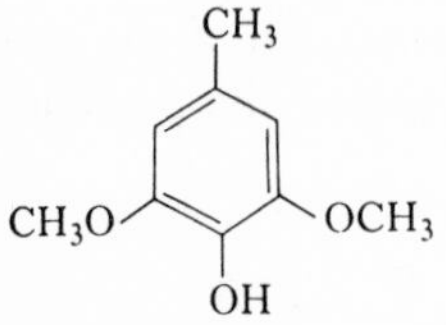

CH₃ / CH₃O— —OCH₃ / OH — (XXXV)

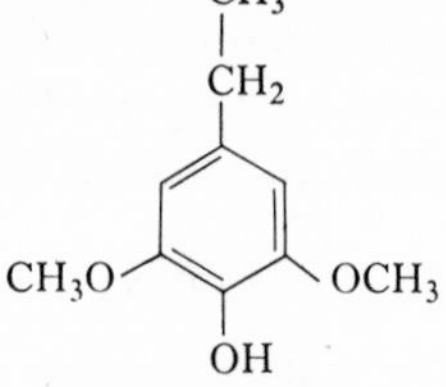

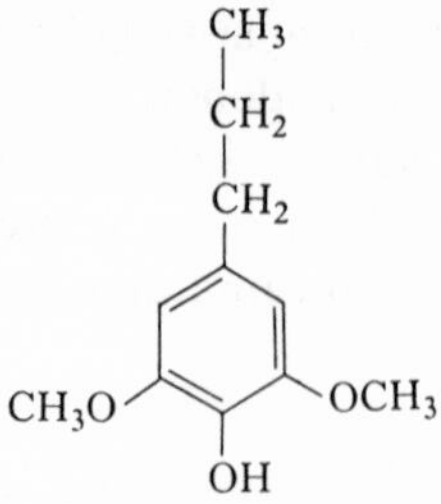

CH₃ / CH₂ / CH₃O— —OCH₃ / OH — (XXXVI)

CH₃ / CH₂ / CH₂ / CH₃O— —OCH₃ / OH — (XXXVII)

CH₂OH / CH₂ / CH₂ / CH₃O— —OCH₃ / OH — (XXXVIII)

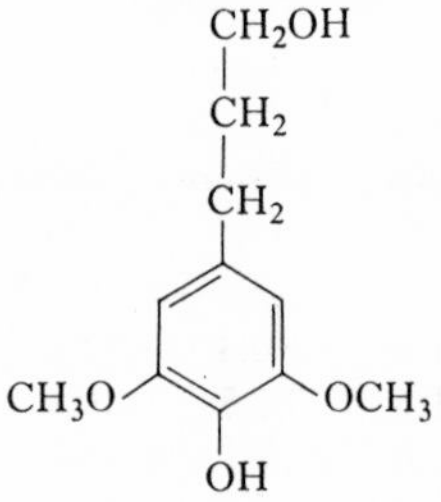

TABLE XII

YIELDS AND MOLAR RATIOS OF HARD-WOOD
LIGNIN HYDROGENATION PRODUCTS (16)

Compounds	Birch			Oak		
	Distill-able (%)	Lignin (%)	Molar ratio[a]	Distill-able (%)	Lignin (%)	Molar ratio[a]
4-Methylguaiacol	2.8	1.1	1.4	3.7	1.0	1.9
4-Ethylguaiacol	2.3	0.9	1.0	2.2	0.6	1.0
4-n-Propylguaiacol	5.8	2.3	2.4	9.5	2.6	4.1
4-Methylsyringol	5.1	2.0	2.1	11.4	3.1	4.8
4-Ethylsyringol	2.8	1.1	1.0	2.6	0.7	1.0
4-n-Propylsyringol	9.8	3.9	3.4	26.9	7.3	9.8
Dihydroconiferyl alcohol	5.1	2.0	1.9	2.1	0.6	0.8
Dihydrosinapyl alcohol	20.0	7.9	6.4	2.9	0.8	1.0
	53.7	21.2		61.3	16.7	

[a]Molar ratios: 4-ethylsyringol assigned value of unity.

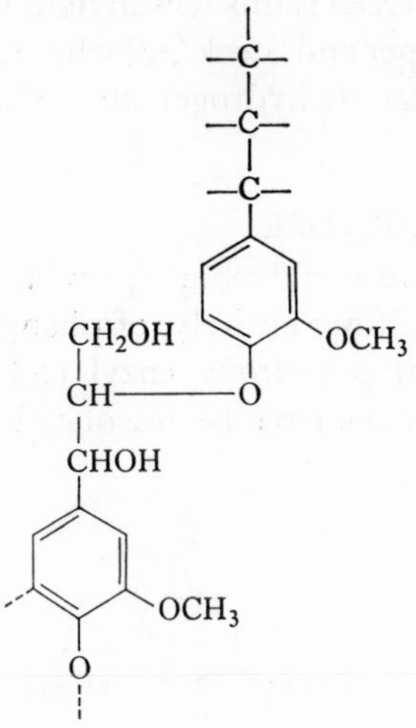

(XXXIX)

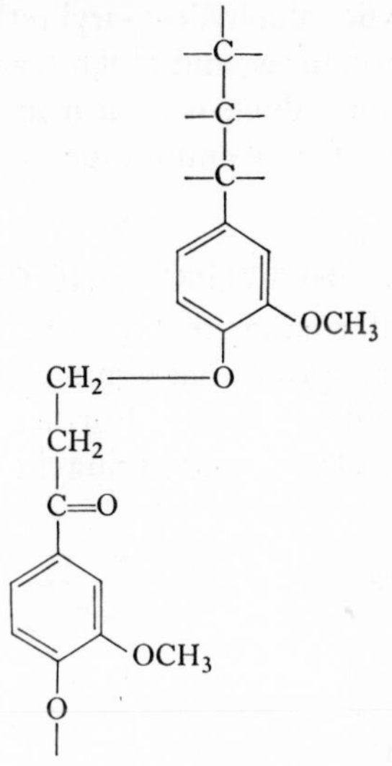

(XL)

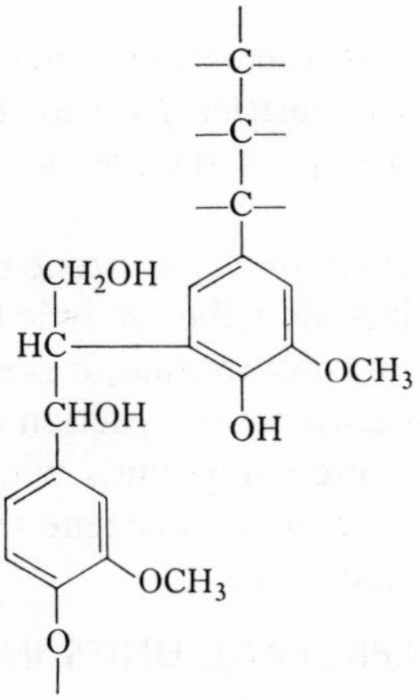

(XLI)

(XLII)

were then subjected to copper chromite-catalyzed high-pressure hydrogenation under the same conditions that had been applied to the milled-wood lignins themselves (38). An interpretation of the results in terms of the hydrogenolysis of carbon–oxygen and carbon–carbon bonds provided evidence for the presence of arylglycerol-β-aryl ether (XXXIX), α-arylhydracrylic aldehyde-γ-aryl ether (XL), and "open" phenylcoumaran (XLI) structures, and of β,β'-carbon–carbon (XLII) linkages in lignin (38).

Similar products were also obtained by Pepper and Steck (42) who studied the effect of time and temperature on the catalytic hydrogenation of aspen lignin.

VIII. THE PHENYLPROPANE CARBON STRUCTURE

The information that has been reviewed above would seem to provide more than ample proof that lignin consists mainly, if not entirely, of phenylpropane building stones. Further the fact that p-hydroxybenzyl (XLIII), vanillyl (XLIV), and syringyl (XLV) derivatives have been isolated from

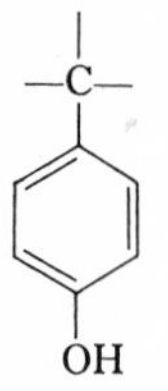

(XLIII) 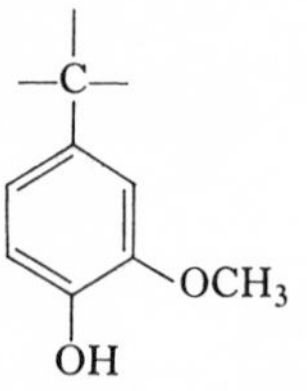   (XLIV) 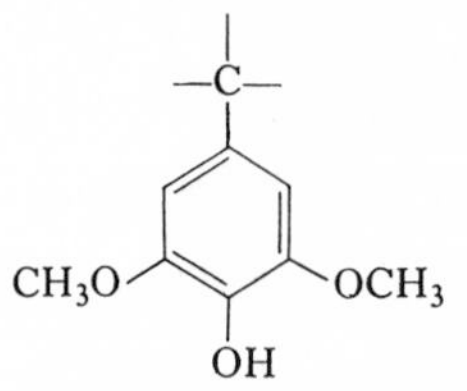   (XLV)

various types of lignin would seem to indicate that the benzene ring carries a hydroxyl group (or etherified hydroxyl) at the position *para* to the side chain, and either zero, one, or two methoxyl groups at the positions *ortho* to the phenolic oxygen function (52).

Convincing proof of the nature of the side chain on the benzene rings is the isolation of the cyclohexylpropane derivatives after the catalytic hydrogenolysis of isolated lignins. From the hydrogen consumption, it is evident that the cyclohexyl rings are the result of a hydrogenation reaction of aromatic (i.e., benzene) rings. The isolation of these compounds with their n-propyl side chains proved that a large number of the benzene rings in lignin also had a three-carbon, i.e., n-propyl, side chain.

IX. MODES OF COMBINATION OF PHENYLPROPANE UNITS IN LIGNIN

There seems to be general agreement that, in the lignin macromolecule, the monomeric phenylpropane units are joined together by both ethereal linkages and by carbon-to-carbon bonds. The carbon-to-carbon bonds are highly resistant to chemical degradation, and constitute the main factor retarding the conversion of the lignins to monomeric units during reactions of ethanolysis, hydrogenation, etc. (46).

If we call the benzene ring of the phenylpropane unit its "head," and the propyl side chain, its "tail," then the units may be said to be joined either "head-to-head," "head-to-tail," or "tail-to-tail." One head-to-head type is a biphenyl-type linkage, in which two benzene rings are joined via a 5–5' bond (XLVI). Tail-to-tail linkages might involve an α–α' combination (XLVII) or a β–β' link (XLVIII). A head-to-tail linkage would involve a β-5' combination (XLIX).

Ethereal linkages may unite phenylpropane units either at just one point, or at more than one. The simpler one-point combinations would include an α-alkyl ether (L) or a β-4'-ethereal linkage (LI). The β-4' ether linkage is typified in the widely recognized guaiacylglycerol-β-aryl ether

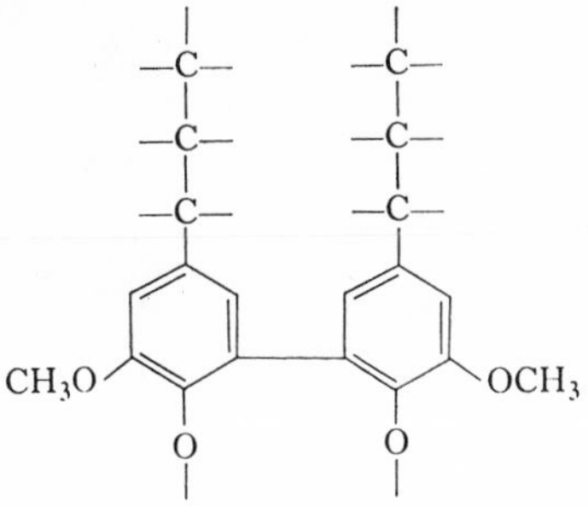

(XLVI)

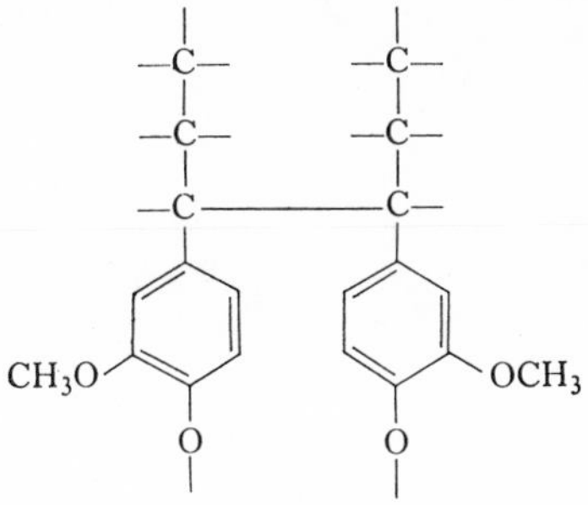

(XLVII)

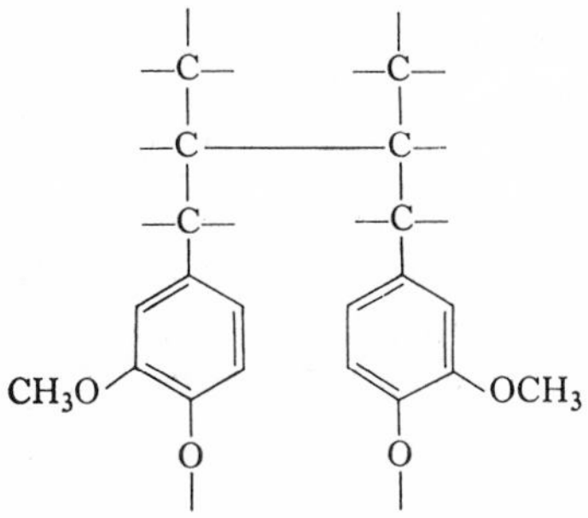

(XLVIII)

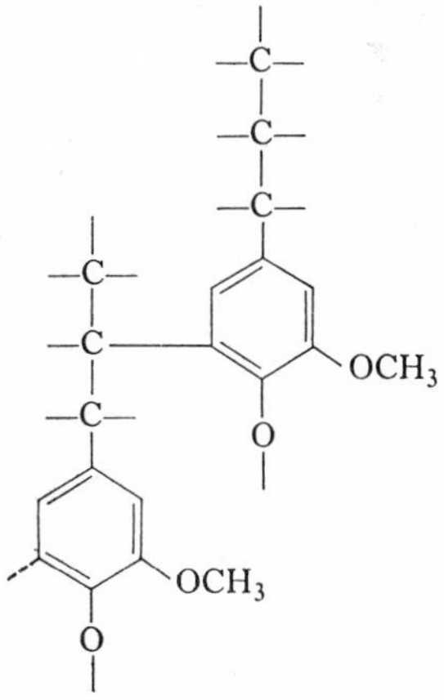

(XLIX)

(L)

(LI)

(LII)

(LIII)

(LIV)

structure (LII) present in lignin (23, 24). Multiple points of attachment of phenylpropane units involving both ethereal *and* carbon-to-carbon linkages are those found in the "benzofuran" (LIII) and in the "pinoresinol" (LIV) type structures, also believed to be present in lignin.

It must still be emphasized that rigorous chemical proof for the existence of many of these linkages in lignin is still lacking. They were originally suggested to account for certain aspects of the chemical behaviour of the lignins. But, the largely speculative original proposals have recently gained some experimental support from studies of the biochemical mechanism of lignin formation and of its degradation in living systems. The chemical evidence supporting these intermonomeric bonds has been reviewed by Sarkanen (46).

REFERENCES

1. H. Adkins, R. L. Frank, and E. S. Bloom, *J. Am. Chem. Soc.* **63**, 549 (1941).
2. E. Adler, K. J. Björkqvist, and S. Häggroth, *Acta Chem. Scand.* **2**, 93 (1948); E. Adler and L. Ellmer, *ibid.* **2**, 839 (1948); E. Adler and S. Häggroth *ibid.* **3**, 86 (1949).
3. A. B. Anderson, *Proc. World Forestry Congr. 5th 1960*, Vol. III, University of Washington, Seattle, Washington, p. 1390, 1960.
4. A. Bjørkman, *Nature* **174**, 1057 (1954).
5. A. Bjørkman, *Svensk Papperstid.* **59**, 477 (1956).
6. J. R. Bower, J. L. McCarthy, and H. Hibbert, *J. Am. Chem. Soc.* **63**, 3066 (1941).
7a. F. E. Brauns, *J. Am. Chem. Soc.* **61**, 2120 (1939).
7b. F. E. Brauns, *Paper Trade J.* **108**, 42 (1939).
8. F. E. Brauns, *J. Org. Chem.* **10**, 211 (1945).
9. F. E. Brauns, "*The Chemistry of Lignin.*" Academic Press, New York, 1952.
10. F. E. Brauns and D. A. Brauns, "The Chemistry of Lignin: Supplement Volume." Academic Press, New York, 1960.
11. F. E. Brauns and H. Seiler, *Tappi* **35**, 67 (1952).
12. C. P. Brewer, L. M. Cooke, and H. Hibbert, *J. Am. Chem. Soc.* **70**, 57 (1948).
13. M. A. Buchanan, F. E. Brauns, and R. L. Leaf, *J. Am. Chem. Soc.* **71**, 1297 (1949).
14. I. K. Buckland, G. H. Tomlinson, and H. Hibbert, *J. Am. Chem. Soc.* **59**, 597 (1937); *Can. J. Res.* **16B**, 54 (1938).
15. C. J. Coscia, W. J. Schubert, and F. F. Nord, *Tappi* **44**, 360 (1961).
16. C. J. Coscia, W. J. Schubert, and F. F. Nord, *J. Org. Chem.* **26**, 5085 (1961).
17. R. Falck, *Ber. Deut. Botan. Ges.* **44**, 652 (1927).
18a. K. Freudenberg, *Angew. Chem.* **68**, 84, 508 (1956).
18. K. Freudenberg, H. Zocher, and W. Dürr, *Ber. Deut. Chem. Ges.* **62**, 1814 (1929).
19. R. E. Glading, *Paper Trade J.* **111**, 32 (1940).
20. E. E. Harris, J. D'Ianni, and H. Adkins, *J. Am. Chem. Soc.* **60**, 1467 (1938).
21. L. M. Cooke, J. L. McCarthy, and H. Hibbert, *J. Am. Chem. Soc.* **63**, 3052, 3056 (1941).
22. H. Ishikawa, W. J. Schubert, and F. F. Nord, *Arch. Biochem. Biophys.* **100**, 131 (1963).
23. H. Ishikawa, W. J. Schubert, and F. F. Nord, *Arch. Biochem. Biophys.* **100**, 140 (1963).
24. H. Ishikawa, W. J. Schubert, and F. F. Nord, *Biochem. Z.* **338**, 153 (1963).
25. E. J. Jones, *Tappi* **32**, 167 (1949).

38   I. INTRODUCTION: THE CHEMISTRY OF LIGNIN

26. L. Kalb and T. Lieser, *Ber. Deut. Chem. Ges.* **61**, 1007 (1928).
27. P. Klason, *Svensk Kem. Tidskr.* **9**, 135 (1897).
28. K. Kratzl and G. Billek, *Holzforschung* **10**, 161 (1956).
29. K. Kratzl and G. Hofbauer, *Monatsh. Chem.* **87**, 617 (1956); **88**, 776 (1957).
30. K. Kratzl and I. Keller, *Monatsh. Chem.* **83**, 205 (1952).
31. K. Kratzl and E. Klein, *Monatsh. Chem.* **86**, 847 (1955); *Mikrochim. Acta*, **1**, 159 (1956).
32. S. F. Kudzin and F. F. Nord, *J. Am. Chem. Soc.* **73**, 690 (1951).
33. S. F. Kudzin and F. F. Nord, *J. Am. Chem. Soc.* **73**, 4619 (1951).
34. H. Mikawa, *Kogyo Kagaku Zasshi*, **54**, 150 (1951).
35. F. F. Nord and W. J. Schubert, *Holzforschung* **5**, 1 (1951).
36. F. F. Nord and G. deStevens, *in* "Handbuch der Pflanzenphysiologie" (W. Ruhland, ed.), Vol. X, p. 389. Springer, Berlin, 1958.
37. A. Olcay, *J. Org. Chem.* **27**, 1783 (1962).
38. A. Olcay, *Holzforschung*, **17**, 105 (1963).
39. A. Payen, *Compt. Rend.* **7**, 1052 (1838).
40. J. M. Pepper, C. J. Brounstein, and D. A. Shearer, *J. Am. Chem. Soc.* **73**, 3316 (1951).
41. J. M. Pepper and H. Hibbert, *J. Am. Chem. Soc.* **70**, 67 (1948).
42. J. M. Pepper and W. Steck, *Can. J. Chem.* **41**, 2867 (1963).
43. D. E. Read and C. B. Purves, *J. Am. Chem. Soc.* **74**, 116, 120 (1952).
44. H. Richtzenhain, *Acta Chem. Scand.* **4**, 206, 589 (1950).
45. H. Richtzenhain, *Chem. Ber.* **83**, 488 (1950).
46. K. V. Sarkanen, *in* "The Chemistry of Wood" (B. L. Browning, ed.), p. 249. Wiley (Interscience), New York, 1963.
47. W. J. Schubert, Doctoral Dissertation, Fordham University, New York, p. 43, 1950.
48. W. J. Schubert, *Holz Roh-Werkstoff* **12**, 373 (1954).
49. W. J. Schubert and F. F. Nord, *J. Am. Chem. Soc.* **72**, 977 (1950).
50. W. J. Schubert and F. F. Nord, *J. Am. Chem. Soc.* **72**, 3835 (1950).
51. W. J. Schubert and F. F. Nord, *Proc. Natl. Acad. Sci. U.S.* **41**, 122 (1955).
52. W. J. Schubert and F. F. Nord, *Advan. in Enzymol.* **18**, 349 (1957).
53. W. J. Schubert and F. F. Nord, unpublished observations (1954).
54. W. J. Schubert, A. Passannante, G. deStevens, M. Bier, and F. F. Nord, *J. Am. Chem. Soc.* **75**, 1869 (1953).
55. F. Schulze, *Chem. Zentr.* **28**, 321 (1857).
56. E. C. Sherrard and E. E. Harris, *Ind. Eng. Chem.* **24**, 103 (1932).
57. P. Shorygin and S. A. Skoblinskaya, *Compt. Rend. Acad. Sci. U.R.S.S.* **14**, 505, 509 (1937).
58. N. N. Shorygina and T. Y. Kefeli, *Zh. Obshch. Khim.* **18**, 528 (1948); **20**, 1199 (1950).
59. H. Staudinger, "Makromolekulare Chemie und Biologie", Wepf, Basel, 1947.
60. G. deStevens and F. F. Nord, *J. Am. Chem. Soc.* **73**, 4622 (1951).
61. G. deStevens and F. F. Nord, *J. Am. Chem. Soc.* **74**, 3326 (1952).
62. G. deStevens and F. F. Nord, *J. Am. Chem. Soc.* **74**, 3447 (1952).
63. G. deStevens and F. F. Nord, *Proc. Natl. Acad. Sci. U.S.* **39**, 80 (1953).
64. J. E. Stone and M. J. Blundell, *Anal. Chem.* **23**, 771 (1951).
65. W. J. Wald, P. F. Ritchie, and C. B. Purves, *J. Am. Chem. Soc.* **69**, 1371 (1947).
66. K. Wiechert, *Cellulosechemie* **18**, 57 (1940).

Chapter II • Aromatization in Microorganisms

I. INTRODUCTION

The mechanism of the biogenesis of lignin has intrigued lignin chemists ever since the discovery of this complex material (3, 4). Simply stated, the process of lignification is the transformation that occurs in certain plants as a result of which the aromatic polymer, lignin, is ultimately synthesized from carbon dioxide, presumably by way of intermediates of a carbohydrate nature. Accordingly, then, lignification is but one illustration of the more general phenomenon of aromatization, i.e., the conversion by living plant cells of non-aromatic precursors into compounds containing benzenoid-type rings. Fundamentally, the problem of the biogenesis of lignin is the elucidation of the enzymic pathway by which this aromatic compound of high degree of polymerization is formed from simpler substances preexisting in the plant.

Although knowledge of the total scheme of biogenesis of lignin is still incomplete, there can be little doubt that it originates ultimately from the carbohydrates which are formed from atmospheric carbon dioxide by the process of photosynthesis. The "lignification problem" may then be considered to include an elucidation of the identity of the ultimate carbohydrate precursors of lignin, together with an understanding of the enzymic mechanisms that are operative, and the knowledge of the identities of the intermediate compounds that are formed by way of which the carbohydrate precursors are eventually transformed into lignin. Obviously, this transformation cannot occur by a direct conversion, but must proceed by way of the polymerization of some simpler, monomeric unit (or units), which are referred to as the "primary building stones." But the almost complete disparity of chemical nature between a carbohydrate, on one hand, and an aromatic polymer, on the other, clearly indicates that an extended series of enzymic reactions is needed to effect this transformation (18, 19).

However, the complexity of the structure of lignin seems to preclude the possibility of the existence of only one simple building unit for lignin in the sense in which glucose or cellobiose are considered the building units of cellulose. Accordingly, for lignin formation, it is also necessary to postulate the existence of several dimeric "secondary building stones." The formation of lignin itself finally may involve either the direct polymerization of the secondary building stones as such, or else, conceivably, there might occur additional modifications before the final polymerization results in the formation of the complex product, lignin.

II. MECHANISM OF THE AROMATIZATION PROCESS IN MICROORGANISMS

Biochemists had speculated for some time about the mechanism by which plants and microorganisms are in general able to achieve the synthesis of benzenoid compounds from nonaromatic precursors. The early investigation of the biogenesis of the aromatic lignin building stones, and of lignin itself, in higher plants had met with much experimental difficulty. However, the opportunity for an experimental approach to the general problem of "aromatization" arose from the isolation of aromatic polyauxotrophs of the microorganisms *Escherichia coli* (6) and *Neurospora crassa* (27), i.e., mutants of these microorganisms that required supplementary mixtures of aromatic compounds in order to conduct their normal metabolic activities.

A. Aromatic Amino Acid Biosynthesis in Bacteria

The research of Davis (7), Sprinson (25), and their collaborators on the mutants of *E. coli* have established a partial pathway for the biosynthesis of the aromatic amino acids from carbohydrate precursors. This work has been reviewed (8, 11b, 13, 25), and the generally accepted pathway for the biogensis of phenylalanine and of tyrosine is presented in Fig. 1.

Thus, the carbon atoms of the two aromatic amino acids are derived from 1 mole of D-erythrose-4-phosphate (II) and 2 moles of phosphoenol-pyruvate (I). The second mole of phosphoenolpyruvate is incorporated by a reaction with 5-phosphoshikimic acid (III) to yield a compound (IV) which forms prephenic acid (V). Prephenic acid then serves as a "branching point"; thus, it may be converted either to phenylpyruvic acid (VI), or to p-hydroxyphenylpyruvic acid (VIII), and these compounds yield phenyl-alanine (VII) and tyrosine (IX), respectively, by transamination reactions.

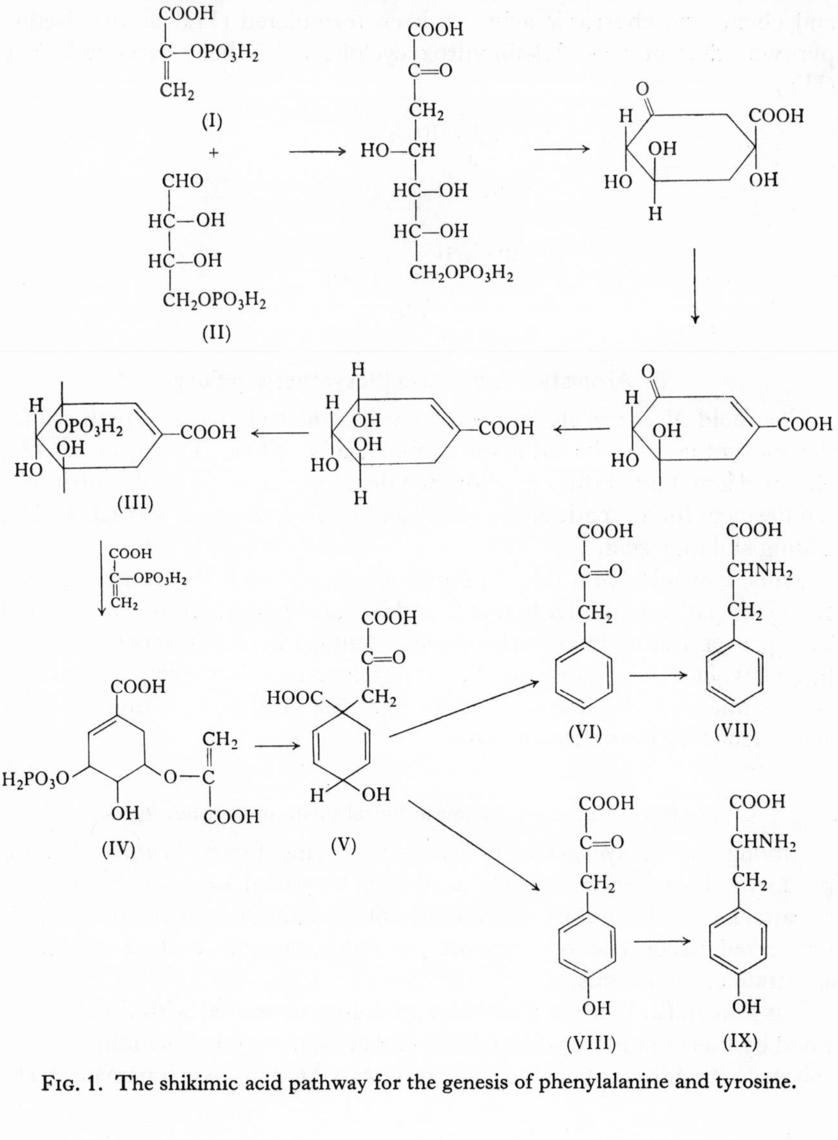

FIG. 1. The shikimic acid pathway for the genesis of phenylalanine and tyrosine.

A new intermediate in aromatic biosynthesis has recently been discovered (11a). This compound, for which the trivial name chorismic acid was suggested, lies in the metabolic sequence after 3-enolpyruvylshikimic acid 5-phosphate (IV) and can be converted enzymically to prephenic acid (V). While further work may be necessary to provide a final proof of its structure and chemistry, chorismic acid has been formulated (11a) as the 3-enolpyruvic ether of *trans*-3,4-dihydroxycyclohexa-1,5-diene carboxylic acid (IVa).

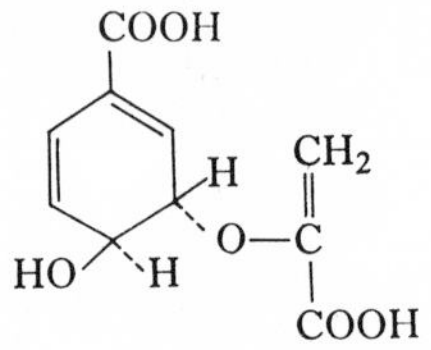

$$
\begin{array}{c}
\text{COOH} \\
\end{array}
$$

(IVa)

B. Aromatic Amino Acid Biosynthesis in Fungi

The mold *N. crassa* also appears to utilize the shikimic acid pathway for the biosynthesis of its aromatic amino acids. Thus, Tatum *et al.* (27) obtained a mutant of this organism that demonstrated a multiple nutritional requirement for aromatic amino acids, and this requirement was satisfied by adding shikimic acid.

Thus, it would seem that *N. crassa* also synthesizes its aromatic amino acids by a pathway similar to that found in *E. coli*, although, as yet, it has not been proven that the two pathways are identical in all respects (13). Other fungi have not been investigated to any great extent. However, the addition of shikimic acid was found (1) to increase the yield of 6-methyl salicyclic acid formed by *Penicillium patulum*.

C. Methyl p-Methoxycinnamate Metabolism in *Lentinus lepideus*

Among the many species of wood-destroying fungi, *Lentinus lepideus* produces "brown rot" in wood, i.e., during its growth on wood, a preferential attack is made on the carbohydrate components and the lignin is largely unaffected, in contrast to "white rot", in which the lignin seems to be a main substrate of the fungus.

It is known (2) that the metabolic processes associated with the decay of wood by this organism give rise to several aromatic esters, including methyl anisate (X), methyl cinnamate (XI), and methyl p-methoxycinnamate (XII).

When *L. lepideus* is cultivated under laboratory conditions, the latter compound predominates and appears as a crystalline deposit in the culture flasks (16).

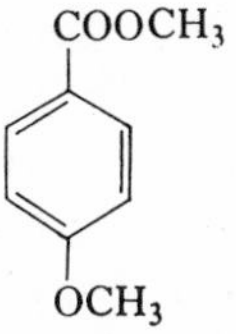

$$
\begin{array}{ccc}
\text{COOCH}_3 & \text{COOCH}_3 & \text{COOCH}_3 \\
| & | & | \\
\text{CH} & \text{CH} & \text{CH} \\
\parallel & \parallel & \parallel \\
\text{CH} & \text{CH} & \text{CH}
\end{array}
$$

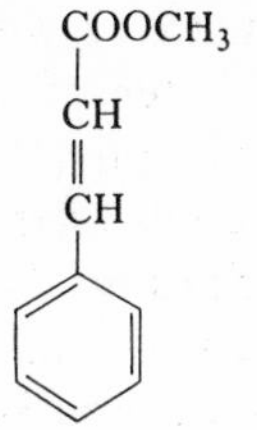

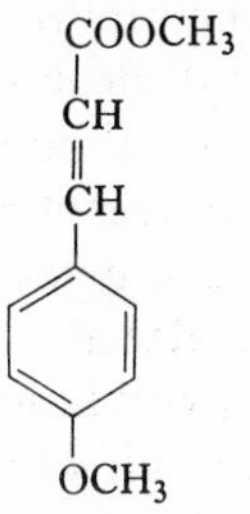

$$
\begin{array}{ccc}
\text{OCH}_3 & & \text{OCH}_3 \\
(\text{X}) & (\text{XI}) & (\text{XII})
\end{array}
$$

Furthermore, it has also been observed (16) that if such growing cultures of *L. lepideus* are allowed to incubate in the presence of the produced crystalline methyl *p*-methoxycinnamate, after a sufficient period of time, the crystals eventually disappear. This indicates a further metabolism of the ester by the fungus (17).

1. Biogenesis of Methyl *p*-Methoxycinnamate

It is possible to grow *L. lipideus* on media containing glucose, xylose, or ethyl alcohol as sole carbon source (16). Methyl *p*-methoxycinnamate then appears as a crystalline deposit in the culture medium after several weeks of growth. From this observation, it was concluded that the ester is not a product of the degradation of lignin (which might conceivably have been effected by the organism during its growth on wood) for the fungus is capable of synthesizing the aromatic ester either from carbohydrates or ethyl alcohol.

Such results obtained from experiments on the biogenesis of methyl

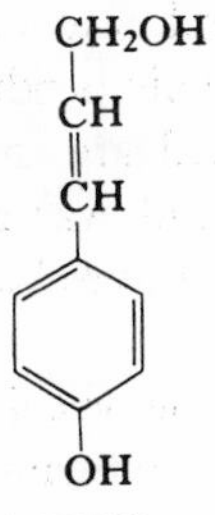

$$
\begin{array}{c}
\text{CH}_2\text{OH} \\
| \\
\text{CH} \\
\parallel \\
\text{CH}
\end{array}
$$

$$
\text{OH}
$$

$$(\text{XIII})$$

p-methoxycinnamate by *L. lepideus* are applicable in theorizing on the formation of lignin building stones (18). The assumption of a similarity in the biogenesis of the ester and of the lignin building stones is based upon the structural relationship of methyl *p*-methoxycinnamate and *p*-hydroxycinnamyl alcohol (XIII), the latter of which is one of three suggested fundamental building stones of lignin (19).

A number of additional products of the metabolism of *L. lepideus* have since been detected (10). Specifically, these are: pyruvic acid, acetoacetic acid, oxalacetic acid, α-ketoglutaric acid, D-ribose, D-glucose, *p*-hydroxyphenylpyruvic acid, sedoheptulose, and 5-phosphoshikimic acid.

The origin of methyl *p*-methoxycinnamate from D-glucose was indicated by the observation that the organism, when grown on ethanol as substrate, synthesized D-glucose, and also, by the results of certain competition experiments. In these experiments, D-ribose, sodium acetate and shikimic acid in the presence of radioactive D-glucose were tested for their ability to serve as competitors of glucose in the biogenesis of methyl *p*-methoxycinnamate. In each case, the ester derived from the competition experiment did not show any dilution of radioactivity when compared with the activity of the product of the control glucose experiment (9).

The detection of the keto acids implied the functioning of the citric acid cycle. Acetic acid can be introduced into this cycle. However, when methyl-C^{14}-labeled sodium acetate was employed, in addition to unlabeled D-glucose, there was no significant incorporation of C^{14} into methyl *p*-methoxycinnamate. This result was interpreted as indicating that neither the keto acids nor acetic acid are directly involved in the formation of methyl *p*-methoxycinnamate by *L. lepideus* (9).

These observations therefore indicated the origin of methyl *p*-methoxycinnamate from D-glucose. This conversion accordingly prompted a comparison with the biogenesis of the aromatic amino acids. Thus, Davis (7) had shown that the synthesis of tyrosine also takes place *via* D-glucose and shikimic acid. Sedoheptulose, shikimic acid, and *p*-hydroxyphenylpyruvic acid appeared among the metabolic products of *L. lepideus*. Shikimic acid and *p*-hydroxyphenylpyruvic acid are intermediates in the biogenesis of tyrosine (7). The latter compound was also identified in the medium of *L. lepideus* cultures, and was considered a precursor of *p*-hydroxycinnamic acid (9).

These findings then indicated the possibility of a relationship between the formation of methyl *p*-methoxycinnamate by *L. lepideus* and the biogenesis of the aromatic amino acids by bacteria (14, 20).

Lentinus lepideus was also grown in media containing D-glucose-1-C^{14} or D-glucose-6-C^{14}. The activities of both tagged sugars were significantly incorporated by *L. lepideus* into methyl *p*-methoxycinnamate. The comparative distributions of radioactivity in the ester derived from the two differently labeled forms of D-glucose were determined (24) by specific degradation reactions which permitted the selective isolation of several of the individual carbon atoms of methyl *p*-methoxycinnamate (XII).

The relative distributions of radioactivity of the ester derived from the two glucose samples are summarized in Table I.

TABLE I

Distribution of Radioactivity in Methyl *p*-Methoxycinnamate Formed from D-Glucose-1-C^{14} and D-Glucose-6-C^{14} (24)

Positions of carbon atoms in methyl *p*-methoxycinnamate	Percentage of total radioactivity of ester	
	From D-glucose-1-C^{14}	From D-glucose-6-C^{14}
1	—	3.2
2, 6	—	39.1
3, 5	—	4.9
4	—	4.2
7	14.1	17.6
8, 9	5.4	5.0
10	14.5	13.4
11	13.2	12.6

The percentage distributions of radioactivity in each carbon of the side chain were nearly identical, in the esters obtained from the 1-C^{14}- and 6-C^{14}-labeled D-glucose; the observed small differences were attributable to a uniformly greater dilution of C-1. In the ester produced from D-glucose-6-C^{14}, significant activity was incoporated into carbons 7 and 2 or 6 of the phenylpropane moiety. In general, these results were similar to those obtained for the tyrosine (4) and shikimic acid (26) biosyntheses from D-glucose.

When the above results were related to those of tyrosine biosynthesis, the possibility that methyl *p*-methoxy-cinnamate was synthesized by *L. lepideus* from D-glucose *via* the shikimic acid pathway was apparent. However, in the ester biogenesis, the specific activity of C-1 underwent a greater dilution than that of C-6. This was accounted for by an alternate oxidative decarboxylation of C-1 of D-glucose (24).

It was also observed that C-6 of D-glucose was significantly incorporated into the methoxy and the ester methyl carbon atoms of the product. The nonequivalent incorporation of carbons 1 and 6 of glucose into these positions gave additional evidence of the operation in *L. lepideus* of a pathway other than glycolysis (24).

These findings then established a relationship between the formation of methyl *p*-methoxycinnamate by *L. lepideus* and the biogenesis of aromatic amino acids by bacteria and fungi. The structural relationship existing between methyl *p*-methoxycinnamate and the lignin building stones has already been discussed. Hence, it seemed possible that the lignin building stones might also be synthesized by a similar pathway (19).

2. METABOLISM OF METHYL *p*-METHOXYCINNAMATE

As indicated previously, *L. lepideus* produces large amounts of crystalline methyl *p*-methoxycinnamate in its culture medium. It has also been observed that if cultures in which the deposit had accumulated are mechanically agitated, the amount of crystalline deposit in the medium diminishes rapidly, and, after a few days, almost completely disappears. Simultaneously, the color of the medium turns brown.

When, after 1 or 2 days of shaking the flasks, the medium was extracted with ether, the presence of a small amount of a previously undetected phenolic compound was observed in the ether extract. This compound was isolated and identified as methyl *p*-coumarate (22). The methyl *p*-coumarate (XIV) was easily oxidized by a mycelial extract of the fungus (22). Hence, it was suggested that methyl *p*-coumarate might function as an intermediate in the metabolism of methyl *p*-methoxycinnamate by *L. lepideus* and possibly also in the biosynthesis of that compound. It was considered that, in the first step of its metabolism, methyl *p*-methoxycinnamate might be demethylated to methyl *p*-coumarate and then, the latter might be oxidized by a phenolase, possibly tyrosinase (22).

In a subsequent investigation (21), the occurrence in the culture medium of the phenolic ester, methyl isoferulate (XV), was also established, together with methyl *p*-coumarate and methyl *p*-methoxycinnamate. It was further noted that methyl *p*-coumarate was accumulated only in small amounts and under certain special conditions, and that it was rapidly oxidized to a colored material by a phenolase present in the fungal mycelium.

When methyl p-coumarate-carboxyl-C^{14} was added to a *L. lepideus* culture medium, about 60% of the total isotopic activity was recovered in the methyl p-methoxycinnamate that was subsequently isolated. The isotopic activity of the carbon in the carboxyl position of methyl p-coumarate did not migrate to other carbons of methyl p-methoxycinnamate. The loss of about 40% of the original isotopic activity was accounted for on the basis of (*a*) mechanical loss during the isolation and purification of the product, (*b*) oxidation of part of the added ester by a phenolase in the fungal mycelium, and (*c*) dilution of the added labeled ester with that synthesized by the fungus. Nevertheless, the significant recovery of the isotopic activity demonstrated that methyl p-coumarate was a precursor of methyl p-methoxycinnamate in the biosynthesis of the latter (21).

In another experiment, 20% of the total activity of added DL-methionine-methyl-C^{14} was recovered from the methyl p-methoxycinnamate subsequently isolated. Here, the isotopic activity was found principally in the ethereal methoxyl and ester methyl carbon atoms. This indicated that methionine (or some related compound) may be the methyl donor for the ethereal and ester methyl groups of methyl p-methoxycinnamate formed by *L. lepideus* (21).

Accordingly, it is believed that methyl p-coumarate is methylated to methyl p-methoxycinnamate by methionine (or related compound). However, the reverse of this reaction might occur under other cultural conditions. Thus, while methyl p-coumarate does not accumulate in the medium (except under special conditions), small amounts of methyl isoferulate do accumulate, along with methyl p-methoxycinnamate.

Two possible explanations for the formation of methyl isoferulate were considered. One is the O-methylation of methyl caffeate (XVI), which could be formed from methyl p-coumarate by the action of a phenolase. The other is the hydroxylation of methyl p-methoxycinnamate by an enzyme other than a phenolase.

Thus, it is believed that in the metabolism of methyl p-methoxycinnamate (XII) and methyl isoferulate (XV), these esters are first demethylated to the free phenolic compounds. (This would correspond to the reverse of their biosynthesis.) Then, the free phenols formed would be subject to the action of a phenolase in the mold mycelium, whereupon colored oxidation products would be formed. These transformations are summarized in Fig. 2.

It is noteworthy that the three phenolic compounds identified were all accumulated in the medium of *L. lepideus* in the form of their methyl esters, and not as free acids. The accumulation of such compounds in culture media is not frequently encountered in studies of microorganisms. Accordingly, the mechanism of the *O*-methylation of phenols in the realm of microbial metabolism cannot yet be fully explained (21).

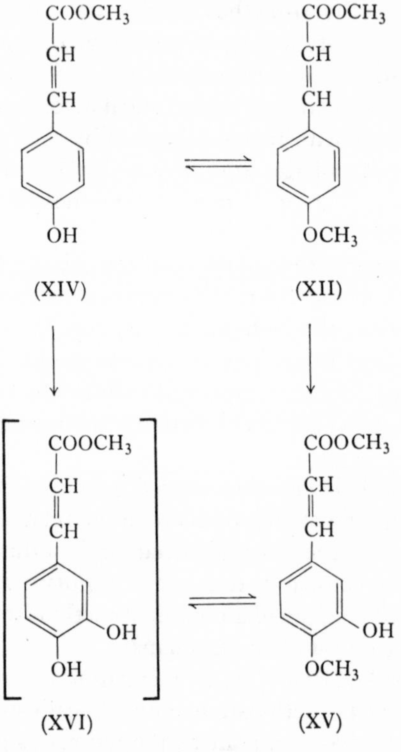

FIG. 2. Transformations in the metabolism of *p*-methoxy aromatic esters by *L. lepideus.*

In a related investigation (23), the transformations of anisic acid (XVII) and methyl anisate (XVIII) by another wood-destroying fungus, *Polystictus*

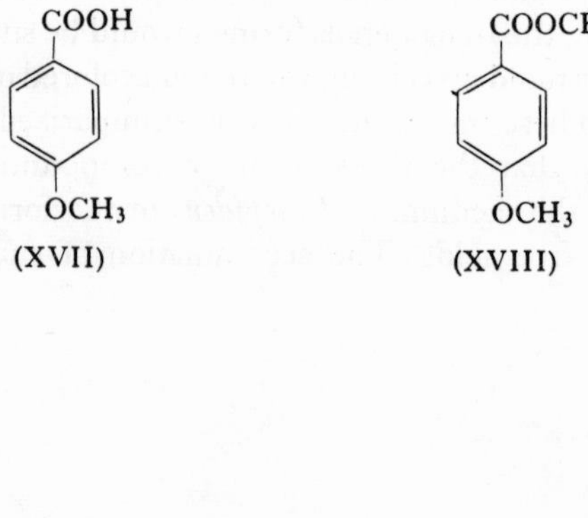

(XVII) (XVIII)

(*Polyporus*) *versicolor*, were studied. In contrast with *L. lepideus*, this organism does not accumulate significant amounts of aromatic compounds in its culture medium. However, when methyl anisate was added to its medium, transformations analogous to those undergone in the methyl *p*-methoxycinnamate metabolism of *L. lepideus* were observed.

Thus, when methyl anisate was added to the medium of *P. versicolor*, demethylation and (simultaneously) hydroxylation of the ester were observed, and the color of the medium gradually turned brown. However, if the cultures were first heated, and then methyl anisate was added to the medium and shaken, no hydroxylated compounds could be detected. On the other hand, the conversion of methyl anisate to hydroxylated compounds was observed after heating the cultures if an ascorbic acid system was also included in the original experiment (23). Accordingly, it was assumed that these transformations were achieved by a similar enzyme system inherently present in this organism.

However, significantly, if free anisic acid, rather than methyl anisate, was added to the medium of this fungus, no hydroxylated derivatives of anisic acid could be detected, and the color of the medium did not turn brown. Instead, the anisic acid was rapidly converted into anisaldehyde (XIX) and anisyl alcohol (XX), both of which were identified in the medium. Thus,

CHO

OCH$_3$

(XIX)

CH$_2$OH

OCH$_3$

(XX)

it was observed that anisic acid was converted to anisaldehyde, and that the major portion of the aldehyde produced was reduced to anisyl alcohol (23).

The observed differences between the transformations of methyl anisate and of anisic acid by *P. versicolor* indicated that the added methyl anisate was not significantly hydrolyzed. However, a slow demethylation and hydroxylation of methyl anisate were achieved by the fungus, and these were followed by the oxidation of a portion of the resultant products. Thus the hydroxylated derivatives of methyl anisate are believed to be partially transformed by the general oxidizing processes of the organism, as a result of which the color of the medium becomes brown (23).

When *p*-hydroxybenzoic acid (XXI) was added to the medium of *P.*

versicolor, *p*-hydroxybenzaldehyde (XXII) was isolated (as its 2,4-dinitro-phenylhydrazone). Hence, it was believed possible that *p*-hydroxybenzoic acid was similarly reduceable to *p*-hydroxybenzaldehyde and to the corresponding alcohol (XXIII) (23).

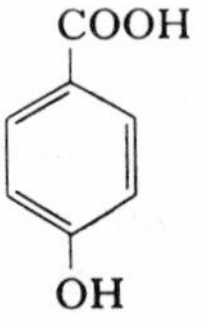

COOH   CHO   CH₂OH

OH   OH   OH
(XXI) 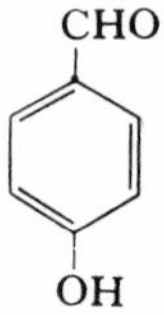   (XXII) 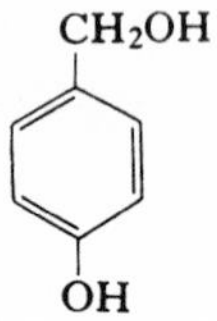   (XXIII)

Evans (11) had also found that the microbiological degradation of benzenoid compounds involves the formation of phenolic compounds at some stage. For example, when soil pseudomonads were grown in a liquid medium containing *trans*-cinnamic acid as sole carbon source, a mixture of phenols was obtained. In this mixture, two phenols were detected; these were identified as melilotic acid (*o*-hydroxyphenylpropionic acid) and 2,3-dihydroxyphenylpropionic acid (5). Hence, it was concluded that these two compounds represented the principal products of the pseudomonad degradation of *trans*-cinnamic acid, prior to ring fission.

Again, the structural relationships existing among products of the metabolism of *L. lepideus*, namely, methyl *p*-methoxycinnamate (XII), methyl *p*-coumarate (XIV), and methyl isoferulate (XV), and the lignin building stones, coniferyl alcohol (XXIV), sinapyl alcohol (XXV), and *p*-hydroxy-cinnamyl alcohol (XIII) are noteworthy. Hence, it has long been considered possible (18) that lignin building stones are synthesized by a metabolic pathway similar to that of the biogenesis of the aromatic esters by *L. lepideus*.

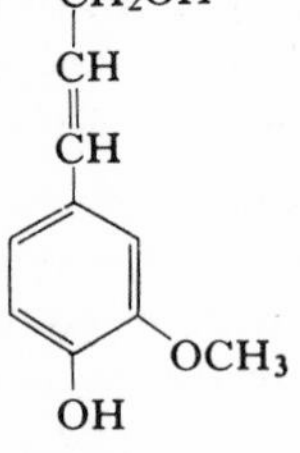

CH₂OH   CH₂OH
CH   CH
CH   CH

              OCH₃   CH₃O   OCH₃
OH   OH
(XXIV) 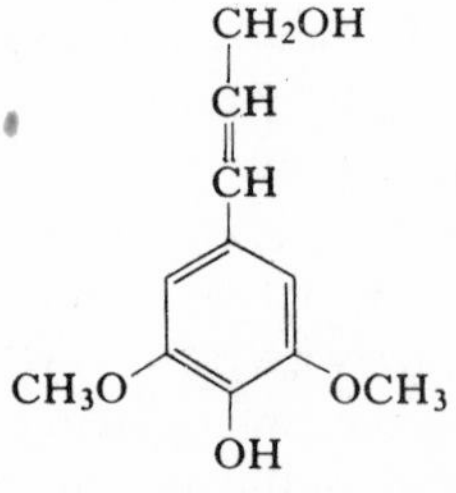   (XXV)

Hence, results obtained in studies of the metabolism, of certain wood-destroying fungi, such as *L. lepideus* and *P. versicolor*, offer a means of investigating the problem of the biogenesis of the lignin building stones. For example, studies on the mechanism of the methylation of phenolic hydroxyl groups by *L. lepideus* (21, 22) could be of significance in relation to the origin of the methoxyl groups of the guaiacyl and syringyl building stones of lignin.

REFERENCES

1. E. W. Bassett and S. W. Tanenbaum, *Biochim. Biophys. Acta* **28**, 247 (1958).
2. J. H. Birkinshaw and W. P. K. Findlay, *Biochem. J.* **34**, 82 (1940).
3. F. E. Brauns, "The Chemistry of Lignin," p. 694. Academic Press, New York, 1952.
4. F. E. Brauns and D. A. Brauns, "The Chemistry of Lignin: Supplement Volume," p. 659. Academic Press, New York, 1960.
5. C. B. Coulson and W. C. Evans, *Chem. Ind. (London)* p. 543 (1959).
6. B. D. Davis, *Experientia* **6**, 41 (1950).
7. B. D. Davis, *Advan. Enzymol.* **16**, 247 (1955).
8. B. D. Davis, *Arch. Biochem. Biophys.* **78**, 497 (1958).
9. G. Eberhardt, *J. Am. Chem. Soc.* **78**, 2832 (1956).
10. G. Eberhardt and F. F. Nord, *Arch. Biochem. Biophys.* **55**, 578 (1955).
11. W. C. Evans, in "Handbuch der Pflanzenphysiologie" (W. Ruhland, ed.), Vol. X p. 454. Springer, Berlin, 1958.
11a. F. Gibson and L. M. Jackman, *Nature* **198**, 388 (1963).
11b. T. Higuchi and I. Kawamura, in "Moderne Methoden der Pflanzenanalyse" (K. Paech and M. V. Tracey, eds.), Vol. VII, p. 260. Springer, Berlin, 1964.
12. J. G. Levin and D. B. Sprinson, *Biochem. Biophys. Res. Commun.* **3**, 157 (1960).
13. A. C. Neish, *Ann. Rev. Plant Physiol.* **11**, 55 (1960).
14. F. F. Nord and W. J. Schubert, *Tappi* **40**, 285 (1957).
15. F. F. Nord and G. deStevens, in "Handbuch der Pflanzenphysiologie" (W. Ruhland, ed.), Vol. X, p. 389. Springer, Berlin, 1958.
16. F. F. Nord and J. C. Vitucci, *Arch. Biochem.* **14**, 243 (1947).
17. F. F. Nord and J. C. Vitucci, *Arch. Biochem.* **15**, 465 (1947).
18. F. F. Nord and J. C. Vitucci, *Advan. Enzymol.* **8**, 253 (1948).
19. W. J. Schubert and F. F. Nord, *Advan. Enzymol.* **18**, 349 (1957).
20. W. J. Schubert and F. F. Nord, *Ind. Eng. Chem.* **49**, 1387 (1957).
21. H. Shimazono, *Arch. Biochem. Biophys.* **83**, 206 (1959).
22. H. Shimazono and F. F. Nord, *Arch. Biochem. Biophys.* **78**, 263 (1958).
23. H. Shimazono and F. F. Nord, *Arch. Biochem. Biophys.* **87**, 140 (1960).
24. H. Shimazono, W. J. Schubert, and F. F. Nord, *J. Am. Chem. Soc.* **80**, 1992 (1958).
25. D. B. Sprinson, *Advan. Carbohydrate Chem.* **15**, 235 (1960).
26. P. R. Srinivasan, M. T. Shigeura, M. Sprecher, D. B. Sprinson, and B. D. Davis, *J. Biol. Chem.* **220**, 477 (1956).
27. E. L. Tatum, S. R. Gross, G. Ehrensvaerd, and L. Garnjobst, *Proc. Natl. Acad. Sci. U.S.* **40**, 271 (1954).

Chapter III • Biogenesis of Lignin in Higher Plants

It would seem appropriate to consider the process of lignification as it occurs in higher plants from the standpoint of the sequence of events that take place in a lignifying plant. The development of mature wood fiber has been divided (78) into four phases: cell division, cell enlargement, cell wall thickening, and lignification. However, these stages are not strictly consecutive and are not to be considered as separate and distinct events.

The biochemical pathway of the lignification phase has been divided (3) into two distinct aspects: (*a*) the formation of the primary lignin building stones, such as p-coumaryl alcohol, coniferyl alcohol, and sinapyl alcohol, and (*b*) the conversion of these building stones into lignin itself.

Although a large number of plant constituents have been postulated as lignin precursors (8, 9), in most cases the evidence supporting these hypotheses has been meager. On the basis of more recent observations, the most tenable theory would appear to be that lignin is a polymer of some compound or compounds with a phenylpropane skeleton, as for example, one or more of the three lignin building stones (62, 66).

Although information on isolated stages of lignin formation has from time to time appeared, the total, continuous, integrated series of biochemical reactions leading ultimately from carbon dioxide, via carbohydrate intermediates, to the primary lignin building stones long remained obscure. The following studies are described in an effort to clarify the pathway by which lignin is synthesized in growing plants.

I. LIGNIN FORMATION FROM CARBON DIOXIDE

After the exposure of sugar cane plants to radioactive carbon dioxide in the dark, the lignin fraction of these plants was found to contain radioactivity (37).

In a study of lignin biosynthesis in wheat plants, Stone *et al.* (75) found that the greatest increase in the production of lignin occurred 45–70 days after seeding. The methoxyl content of the plants was also found to increase as the plants matured. Stone (74) then subjected the wheat plants to $C^{14}O_2$ in a "long-term" experiment, at a stage of growth corresponding to rapid lignification. The plants were harvested every few days until maturity, and were then oxidized with alkaline nitrobenzene, and the resulting vanillin, syringaldehyde, and p-hydroxybenzaldehyde were separated. The results indicated that all the $C^{14}O_2$ was incorporated into the lignin within 24 hours after administration. The total activity originally acquired by the syringaldehyde portion of the lignin remained constant throughout the growth of the plant. From the results, Stone concluded that lignin was an end product of plant growth.

In "short-term" experiments (11), wheat plants, again at a stage of rapid lignification, were exposed for 20 minutes to $C^{14}O_2$ in a closed chamber, and were then grown for 1 to 24 hours in a normal atmosphere before harvesting. Analytical results indicated that the synthesis of lignin was most rapid from 4 to 6 hours after $C^{14}O_2$ administration. After 24 hours, C^{14} appeared in the lignin to the extent of about 1.5–2.0% of that administered.

II. CARBOHYDRATE PRECURSORS

An investigation of the biochemistry of lignification includes an elucidation of the identity of the carbohydrate precursors from which lignin is ultimately derived, as well as the nature of the enzymic reactions operative during the process. As a result of these transformations, the carbohydrate precursors, photosynthetically derived from atmospheric carbon dioxide, are eventually converted into lignin (66).

Many suggestions have been advanced with regard to the identity of the carbohydrate precursor(s), but these suggestions have been justly referred to as either purely speculative or based on evidence of an indirect or fragmentary nature (63).

A. The Role of D-Glucose

In an early investigation (64), uniformly labeled D-glucose was fed to a Norway spruce tree, and, after a period of metabolism, radioactivity was detected in the cambium layer of the tree. The lignin of this layer was isolated and found to be radioactive (Table I).

TABLE I

DISTRIBUTION OF RADIOACTIVITY IN NORWAY SPRUCE TREE AFTER
FEEDING UNIFORMLY LABELED D-GLUCOSE-C^{14} (64)

Plant material	Activity (counts/min./mg. C)
Stem	460
Lignin	240

From the data of Table I, it appeared that the radioactivity of the labeled D-glucose fed to the tree was incorporated to a considerable extent into its lignin. Thus, the Norway spruce tree had the capacity to convert D-glucose into lignin. Considering the central position that D-glucose occupies in plant biochemistry, both as a product of photosynthesis and as the monomeric unit of cellulose, it was considered significant that the Norway spruce tree also possessed the enzymic ability to convert this monosaccharide into lignin (64). Obviously, however, in the case of lignin biogenesis, an aromatization process is also involved. This will be considered in a later section.

In a subsequent investigation (2), the fate in lignification of D-glucose preparations which were labeled with C^{14} specifically in their number 1 or 6 positions was studied. In these experiments, D-glucose-1-C^{14} and D-glucose-6-C^{14} were fed separately to individual Norway spruce trees. The method of feeding the tagged compounds and of the isolation of the resultant lignin have been described (60). The isolated lignin was subjected to an alkaline nitrobenzene oxidation, and vanillin was obtained. Carbons 1, 2, 5, 6, 7, and 8 of this vanillin were isolated as $BaCO_3$ by means of the degradation reactions shown in Fig. 1.

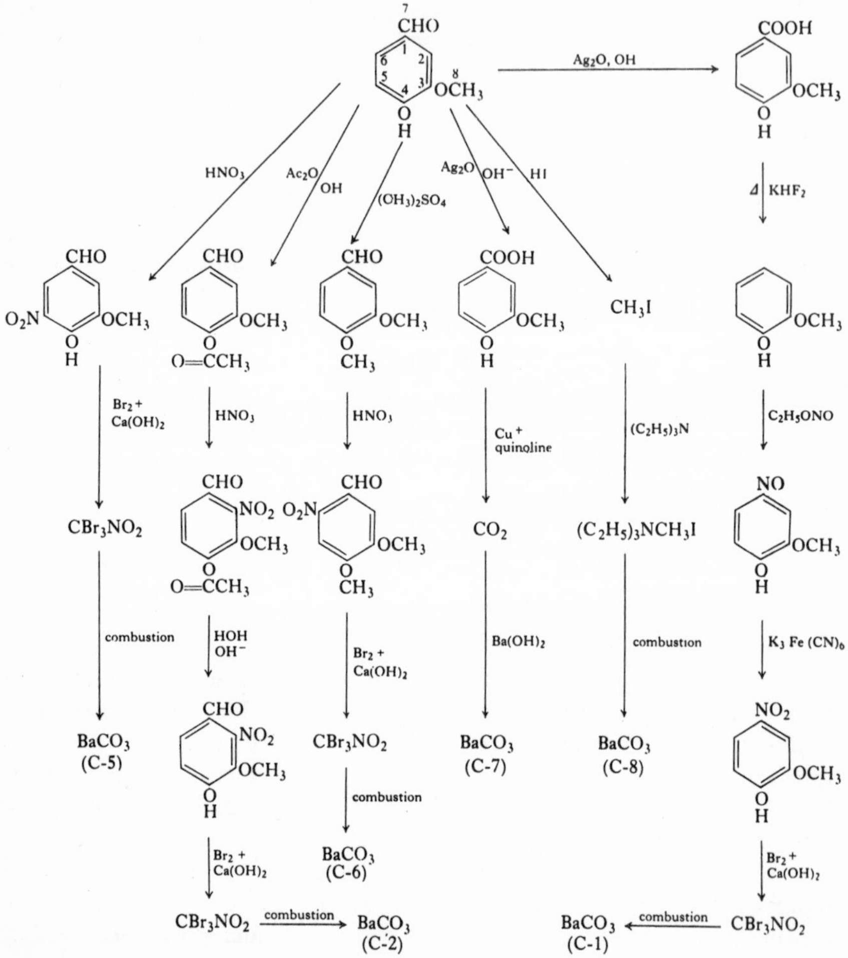

FIG. 1. Degradation reactions resulting in the isolation of carbons 1, 2, 5, 6, 7, and 8 of vanillin.

The radioactivities of the individual compounds obtained after the degradation of the lignins obtained in the two specifically labeled D-glucose experiments are given in Table II.

A comparison of the distribution of radioactivity in the various positions of vanillin from the two experiments shows that appreciable activity was incorporated into carbons 2, 6, 7, and 8, whereas considerably less was

TABLE II

PERCENTAGE DISTRIBUTION OF RADIOACTIVITY IN VARIOUS
POSITIONS OF VANILLIN OBTAINED FROM D-GLUCOSE-1-C^{14}
AND D-GLUCOSE-6-C^{14} EXPERIMENTS (2)

	Percentage of total radioactivity in vanillin	
Positions in vanillin	D-Glucose-1-C^{14} experiment	D-Glucose-6-C^{14} experiment
Total molecule	100.0	100.0
C-1	2.6	4.5
C-2	18.2	16.1
C-5	3.1	3.9
C-6	11.4	18.1
C-7	28.5	22.0
C-8	31.1	24.7
C-3, 4[a]	5.1	10.7

[a] Calculated values.

incorporated into the other positions. These results are therefore similar to those obtained for the biosyntheses of shikimic acid (71) and of methyl *p*-methoxycinnamate (67) from glucose, and have been amply corroborated (47a, b).

It has been asserted (62, 63, 66) that both fungal and plant biosyntheses of phenylpropane moieties might follow similar pathways. The similarity of C^{14} distribution in the aromatic rings of the methyl *p*-methoxycinnamate

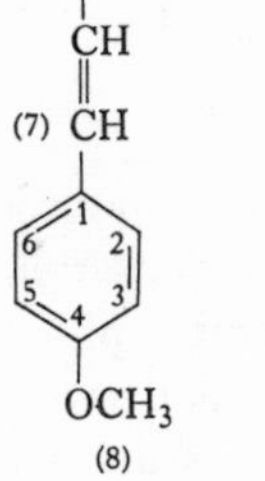

(I) 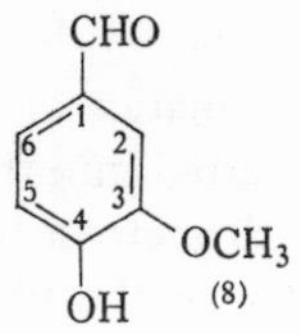   (II)

formed by *L. lepideus* from D-glucose-1-C^{14} and D-glucose-6-C^{14} and of the vanillin derived from the lignin oxidations indicated that a marked similarity in metabolic pathways might indeed exist in these two apparently unrelated systems. This becomes apparent upon comparison of the percentage distributions of C^{14} in the methyl *p*-methoxycinnamate (I) derived from *L. lepideus* and the vanillin (II) derived from the lignin of Norway spruce, using corresponding position designations.

The comparison in Table III reveals a remarkable agreement in the relative distribution of radioactivity in the two experiments. From the data, it is apparent that both the ester and the vanillin incorporated most of the radioactivity in carbons 2, 6, 7, and 8.

TABLE III

COMPARISON OF RELATIVE DISTRIBUTIONS OF RADIOACTIVITY
IN CORRESPONDING POSITIONS OF METHYL *p*-METHOXY-
CINNAMATE AND VANILLIN (60a, b)

Corresponding positions in ester and vanillin	Percentage distribution of C^{14}	
	Methyl *p*-Meth-oxycinnamate	Vanillin (2)
C-1	3.2	4.5
C-2	19.5[a]	16.1
C-3	3.5[a]	5.4[a]
C-4	4.2[a]	5.4[a]
C-5	3.5[a]	3.9
C-6	19.5[a]	18.1
C-7	17.6	22.0
C-8	12.6	24.7

[a] Calculated values.

B. The Cyclization Step

The mode of formation of shikimic acid from D-glucose (71) indicates that the sugar is converted into this acid by the condensation of a triose derived via glycolysis, with a tetrose derived via the pentose phosphate pathway (69). It may be noted that the Embden-Meyerhof pathway produces phospho-enolpyruvate, while D-erythrose-4-phosphate is derived from the pentose phosphate pathway. Significantly, there is evidence that phosphoenolpyruvate and D-erythrose-4-phosphate, implicated in the synthesis of shikimic acid, are formed in higher plants (57).

Evidence for the formation of D-erythrose-4-phosphate in plants rests on its function as a substrate for the enzymes, transketolase and transaldolase, which are believed to form D-erythrose-4-phosphate as an intermediate during respiration by the pentose phosphate pathway (70) and during photosynthesis (14). The formation of D-erythrose-4-phosphate is postulated to explain the cyclic nature of these two processes (57). Furthermore, evidence exists (55) for the formation of D-erythrose-4-phosphate during photosynthesis by *Chlorella*.

These considerations then provide an explanation for the observed radioactivity in positions 2 and 6 of the cyclohexene ring of shikimic acid (71) derived from D-glucose, and of vanillin (2) obtained as described above, as shown in Fig. 2.

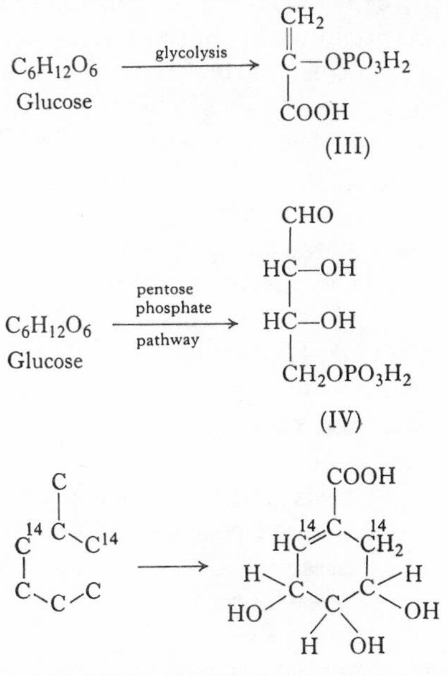

$$C_6H_{12}O_6 \xrightarrow{\text{glycolysis}} \begin{array}{c} CH_2 \\ \parallel \\ C-OPO_3H_2 \\ \mid \\ COOH \end{array}$$

Glucose

(III)

$$C_6H_{12}O_6 \xrightarrow[\text{pathway}]{\substack{\text{pentose} \\ \text{phosphate}}} \begin{array}{c} CHO \\ \mid \\ HC-OH \\ \mid \\ HC-OH \\ \mid \\ CH_2OPO_3H_2 \end{array}$$

Glucose

(IV)

(V)

FIG. 2. Formation of shikimic acid from D-glucose.

Thus, the ultimate organic source of lignin is the carbohydrates photosynthetically formed by the plant from atmospheric carbon dioxide. Phosphoenolpyruvate and D-erythrose-4-phosphate may serve as proximate precursors of the aromatic rings of lignin, while D-glucose was demonstrated (2) to be an ultimate source.

Fully labeled xylitol was also infused into sprucewood, and was found to be incorporated into the lignin as well as the polysaccharides (48). After isolation of the radioactive wood constituents, they were chemically degraded, and the radioactivity distribution was quantitatively determined. In this way, the conversion of xylitol to lignin was demonstrated (48). Apparently, then, D-glucose is not the only carbohydrate which is convertible into lignin.

III. THE SHIKIMIC ACID PATHWAY

As described in the previous chapter, the studies of Davis (17) and of Katagiri (42, 43) established the fact that shikimic acid played an important role in bacterial and fungal metabolism as a precursor of aromatic amino acids, and that the enzyme systems that brought about the transformation of glucose, via shikimic acid and other intermediates, to the aromatic acids were found in microorganisms. Furthermore, the presence of the enzyme system responsible for the synthesis of shikimic acid was confirmed, not only in microorganisms, but also in higher plants (54). These investigations suggested the possibility that other aromatic products, such as lignin, which is widely distributed in higher plants, may also be formed via a reaction sequence similar to the shikimic acid pathway.

Accordingly, in relation to the biosynthesis of lignin, it is of importance to consider the distribution of shikimic acid in higher plants, and particularly in woody plants. Hasegawa et al. (38) investigated the occurrence of shikimic acid in the leaves of one hundred and sixty-four plant species, and found shikimic acid in eighty-two of them. Higuchi (40) studied the distribution of shikimic acid in the leaves and cross sections of the young stems of ninety-six species of woody plants; here, shikimic acid was detected in seventy of the species. Hence, there can be little doubt that the shikimic acid pathway does function in higher plants (40, 57). The compound, shikimic acid, has been known since the late nineteenth century (24 a).

Experiments with Sugar Cane Plants

The following experiments with sugar cane plants indicate that shikimic acid, without any rearrangement of the carbon atoms of its six-membered ring, functions as a precursor of the aromatic rings of the lignin building stones, and accordingly, that the formation of the latter parallels the mode of formation of the aromatic amino acids in microorganisms.

Specifically labeled shikimic acid was prepared by fermentation of D-glucose-6-C^{14} by *Escherichia coli* mutant 83–24. Such shikimic acid (Va) contains 44% of its total activity in position 2, and 52% in position 6 (71).

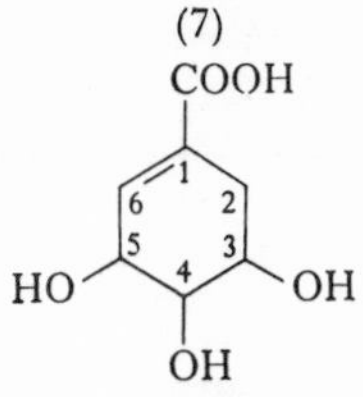

(7)
COOH

HO   OH

OH

(Va)

An aqueous solution of this specifically labeled shikimic acid was incorporated into a growing sugar cane plant. After several days of metabolism, the leaves were removed, the stem of the plant was cut, dried, and pulverized, and the resulting powder was thoroughly extracted with water. Radioactivity measurements of the resultant plant materials (Table IV) indicated that, upon introduction of the labeled shikimic acid into the plant, the radioactivity was incorporated into nonwater-extractable components of the stem. Examination of the isolated Klason lignin revealed that the highest radioactivity was to be found in the lignin.

TABLE IV

DISTRIBUTION OF SHIKIMIC ACID RADIOACTIVITY
IN SUGAR CANE PLANT (20)

Plant material	Radioactivity (counts/min.)
Stem (ground and water-extracted)	6
Klason lignin (10% of weight of stem)	42
Vanillin	58

The pulverized plant material was first submitted to treatment with Schweizer's reagent to remove the cellulose, and then to an alkaline nitrobenzene oxidation, and the resulting vanillin was isolated.

The distribution of radioactivity in the ring carbons of the vanillin was determined after degradation according to reactions analogous to those shown in Fig. 1. The activities in positions 2, 5, and 6 of the ring, determined after such degradations, are shown in Table V.

TABLE V

DISTRIBUTION OF RADIOACTIVITY IN SELECTED POSITIONS OF
VANILLIN (20)

Vanillin	Radioactivity (counts/min.)	Percentage distribution of total radioactivity
Total molecule	58	100
C-2	190	41
C-5	0	0
C-6	204	44

From these data, it may be seen that the distribution of activity in the aromatic ring of vanillin agrees well with the original distribution of C^{14} in the cyclohexene ring of the incorporated shikimic acid (20).

No attempt is made here to equate carbon 2 of vanillin with the corresponding position in shikimic acid, or carbon 6 of vanillin with this position in the acid, since these two positions were indistinguishable, and may possibly be interconvertible.

Thus, after the absorption of specifically labeled shikimic acid into a sugar cane plant, it was apparent that this compound was metabolized by the plant, and was incorporated to a great extent into its lignin. Further, the degradation of the lignin, via vanillin, revealed that the distribution of activity in the aromatic rings of the product was comparable to that in the incorporated shikimic acid. From these results, it was concluded that shikimic acid is an intermediate in the biochemical pathway from carbohydrates, formed from atmospheric CO_2 by photosynthesis, to the aromatic rings of the lignin building stones (59, 60, 60a, b).

IV. GENESIS OF THE LIGNIN BUILDING STONES

The use of the term "primary lignin building stone" is based upon the assumption that lignin, like cellulose and starch, has a chainlike structure, composed of simple "building stones" which, in turn, are linked in some way (or more likely in several ways) to form a "lignin building unit" (8, 9). This would correspond to the concept of "glucose anhydride" as the building stone for cellulose and starch.

The building stones of lignin, however, possess a phenylpropane carbon structure, and obviously several of these are linked together to form the "lignin building unit." A long series of lignin building units then may make up the total lignin polymer. But, unlike the uniform building stones of the polysaccharides, lignin building stones, although they all have the same basic phenylpropane carbon structure, may be of the coniferyl (VI), sinapyl (VII), or p-coumaryl (VIII) types (8, 9, 66).

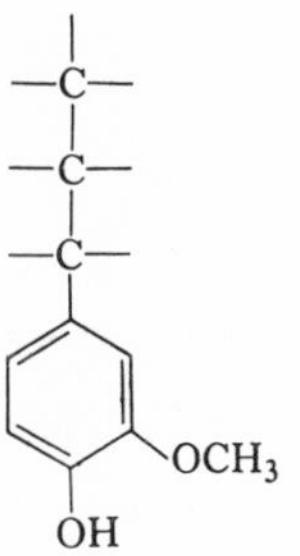

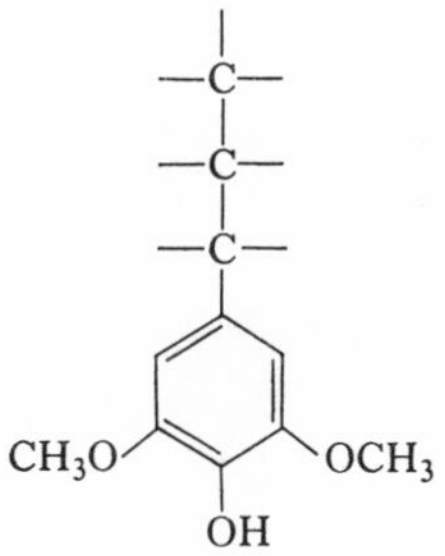

The phenolic cinnamic acids are regarded as potential precursors of lignin since they have the required phenylpropane carbon structure, they are widely distributed in plants (5), and are ionized under physiological conditions (18).

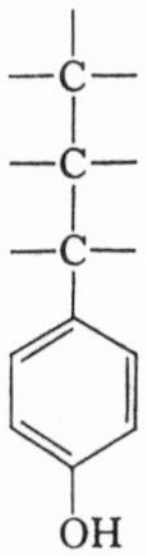

Much speculation about the mechanism of lignification has centered on coniferyl alcohol as the "key" intermediate in a process involving oxidation and polymerization. This idea seems to have developed from three circumstances: (a) the isolation of the glucoside, coniferin, from certain conifers (b) the similarity in elementary analysis of isolated Klason lignin and coniferyl alcohol and (c) the tendency of coniferyl alcohol to polymerize (50).

However, since coniferyl alcohol (IX) itself is not easily detectable in plants, this hypothesis was based on the well-known natural occurrence of coniferin (X), the glucoside of coniferyl alcohol. Similarly, syringin (XII), the glucoside of sinapyl alcohol (XI), is regarded as a precusor of lignin in hardwoods. An earlier survey (49) of the recorded occurrences of coniferin revealed that the presence of this glucoside was established predominantly in coniferous wood species, with six families and fifteen species represented. Except for one discovery of its presence in black locust, syringin has been isolated only from five genera of the olive family (49).

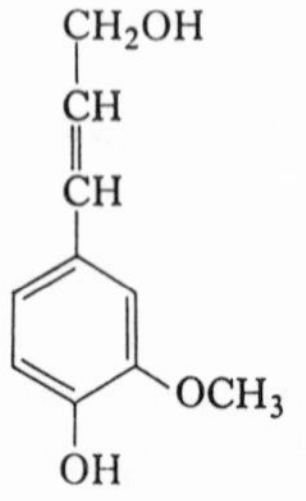

CH₂OH — CH₂OH → HC structures

(IX) 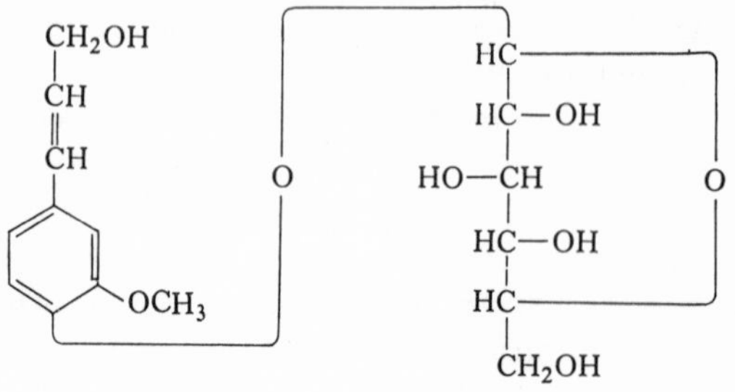   (X)

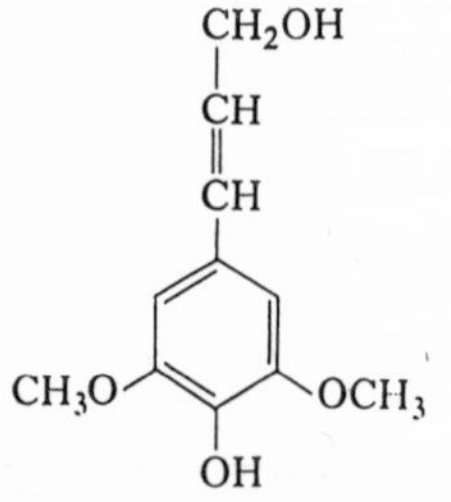

(XI) 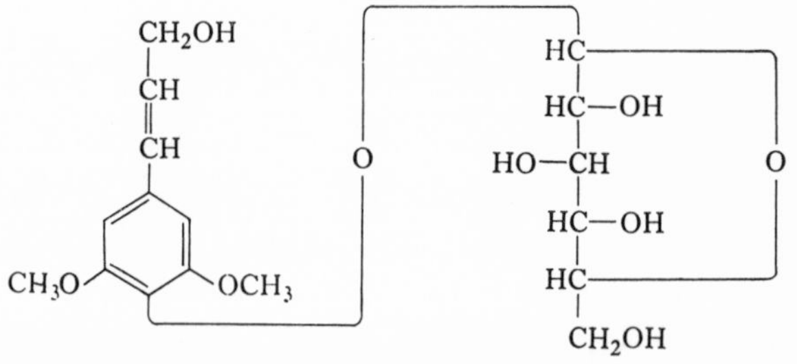   (XII)

The results of tracer studies from several laboratories have now supported the concept that phenolic cinnamic alcohols and acids do function as precursors in the biogenesis of lignin. For example, coniferyl alcohol has been synthesized with radioactive carbon in predetermined positions of the molecule (26, 31, 46), fed to growing plants, and the radioactivity then was detected in the lignins.

Of equal significance has been the elaboration of involved degradation procedures used to recover the radioactivity from the isolated lignin, or, more specifically, from vanillin obtained on the oxidation of lignin (2, 20) or from the so-called "Hibbert ketones" obtained on ethanolysis (46).

The synthesized specifically labeled compounds have been administered to growing plants in various ways. They may be "fed" directly to the plant or to excised shoots (57), or solutions of them may be absorbed through the ends of freshly cut stems, branches, leaves, or needles (2, 20). The plant is then allowed a period of time for metabolism; the lignin is then isolated and its specific activity is measured.

Data obtained in this way have been accepted as proof that suspected precursors, such as coniferyl alcohol, do mediate in lignin biosynthesis (3). Similar studies have also been made with sinapyl and p-coumaryl alcohols.

A large number of related C^{14}-labeled phenylpropane compounds have been studied in this way for their efficiency to serve as precursors of lignin.

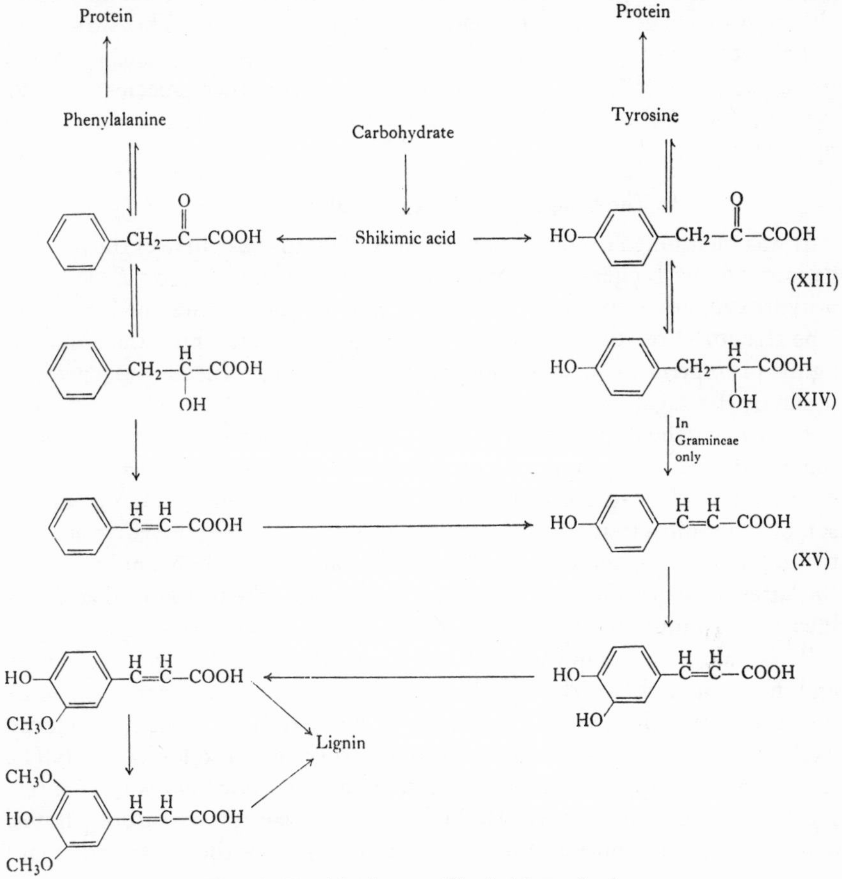

FIG. 3. Partial scheme of lignin biosynthesis.

Experiments of this kind have produced the scheme shown in Fig. 3. It may be noted from this scheme that the cinnamic acid derivatives are obtained from the aromatic amino acids, phenylalanine and tyrosine, and serve as intermediates in the biosynthesis of certain phenolic metabolites peculiar to plants, including lignin (9a, 57).

Many of the compounds shown as intermediates in Fig. 3 were fed by Brown *et al.* to excised shoots of wheat and maple plants, and were found to be readily converted to lignin (10, 12). But in these investigations, certain taxonomic differences were noted. For example, out of eleven species (representing ten plant families), only two converted tyrosine to lignin, although all had utilized phenylalanine as a lignin precursor. The two species utilizing tyrosine were both members of the Gramineae family. Accordingly, it was suggested (12) that the failure of the non-Gramineae species to utilize tyrosine for synthesizing lignin was due to their enzymic inability to dehydrate *p*-hydroxyphenyllactic acid (XIV).

The Role of *p*-Hydroxyphenylpyruvic Acid

It was mentioned in the previous chapter that in the course of the investigations on the biogenesis of methyl *p*-methoxycinnamate by *L. lepideus*, *p*-hydroxyphenylpyruvic acid was detected in the culture medium (19). The structural relationship of this acid to the suggested building stones of lignin prompted an investigation of its possible role in the biogenesis of lignin by the sugar cane plant.

p-Hydroxyphenylpyruvic acid-$C^{14}OOH$ was incorporated into a growing sugar cane plant by employing essentially the same technique as has been described for the D-glucose and shikimic acid experiments (61). A comparison of the radioactivity of the introduced *p*-hydroxyphenylpyruvic acid, of the isolated lignin, and of the barium carbonate obtained on combustion of the latter, revealed that most of the radioactivity of the introduced acid was incorporated into the lignin (65).

This radioactive lignin was subjected to alkaline nitrobenzene oxidation and the vanillin obtained was isolated, and, in this instance, was found to be nonradioactive. However, subjection of the lignin to alkaline fusion produced oxalic acid, which was isolated and found to contain the radioactivity (1). Accordingly, it appeared that *p*-hydroxyphenylpyruvic acid was an intermediate on the pathway between shikimic acid, derived from carbohydrates, and the building stones in the biogenesis of lignin by the sugar cane plant (58, 59).

On the other hand, Billek (7) reported that *p*-hydroxyphenylpyruvic acid is not converted into lignin by spruce trees. This apparent discrepancy was resolved by Neish (56) who first noted that *p*-hydroxyphenylpyruvic acid-3-C^{14} is converted to guaiacyl or syringyl lignin in wheat, but not in buckwheat or sage. This difference in biosynthetic abilities among different species was subsequently interpreted (12) as follows. Neither *p*-hydroxyphenylpyruvic acid (XIII) nor p-hydroxyphenyllactic acid (XIV) is a *general* intermediate in lignification, and certain differences noted (as here) between grasses and nongrasses probably resulted from the unique ability of grasses to convert p-hydroxyphenyllactic acid to p-hydroxycinnamic acid (XV). Since sugar cane plants are classified as grasses, they *are* able to utilize p-hydroxyphenylpyruvic acid for lignin biosynthesis (15).

V. CONVERSIONS OF THE LIGNIN BUILDING STONES

As already described, the hypothesis that lignin is biogenetically derived from coniferyl alcohol is suggested by the presence of the glucoside, coniferin, in precambial tissue (77). The structure of coniferin was elucidated by Tiemann and Mendelsohn in 1875. Shortly, thereafter, they (76) suggested that coniferyl alcohol, the aglucone of coniferin, bore a structural relationship to lignin. Then, in 1899, Klason (44) expressed the opinion that lignin was a condensation or polymerization product of coniferyl alcohol. Later, he suggested (45) that coniferyl aldehyde might be the fundamental building stone of lignin.

But the chemical nature of the polymerization of such monomers remained obscure until 1908 when Cousin and Hérisséy (16) observed that isoeugenol underwent dimerization in the presence of air and under the influence of the oxidase enzymes in a glycerol extract of mushroom. The product of the dimerization, dehydrodiisoeugenol, was subsequently found by Erdtman (21) to have the structure of a phenylcoumaran derivative.

This discovery led Erdtman to suggest a mechanism for the polymerization of monomers into lignin. He postulated (22) that p-hydroxyphenylpropane compounds with unsaturated side chains, on oxidation, underwent coupling reactions in the position *ortho* to the phenolic hydroxyl group, and also at the β-carbon of the side chain. Lignin could then originate from such guaiacylpropane units which are first oxidized in the side chain, and then dehydrogenated. Erdtman concluded that lignin was probably derived from precursors with structures similar to coniferyl alcohol (22). Subsequent to these hypotheses, he now considers (24) that coniferous lignin is derived principally from coniferyl alcohol.

Manskaja (51) investigated the chemical reactions that take place in cambial tissue. Her results led her to believe that lignification in plants takes place by enzymic oxidation-reduction reactions that occur during the course of the metabolic conversions of coniferyl alcohol, which is thereby oxidized, and is eventually converted into the polymerized lignin (52).

About 10 years after the original suggestions of Erdtman (22), these concepts were adopted by Freudenberg (25), who found that D-coniferin, labeled with C^{14} in its side chain, when incorporated into a spruce twig, was deposited in the stem in the form of a high molecular weight material. Ethanolysis of this product yielded radioactive Hibbert ketones. This indi-cated that D-coniferin was hydrolyzed by a β-glucosidase; the liberated coni-feryl alcohol then polymerized to a ligninlike material. Accordingly, Freudenberg suggested the following over-all scheme for the formation of lignin:

$$\text{D-Coniferin} \xrightarrow{\text{β-glucosidase}} \text{coniferyl alcohol} \xrightarrow{\text{phenol dehydrogenase}} \text{lignin}$$

Freudenberg then attempted to obtain polymeric products with ligninlike properties by the *in vitro* treatment of "primary lignin building stones" with air in the presence of a mushroom oxidase (32). Of the compounds tested, the dehydrogenation product of coniferyl alcohol (so-called DHP) showed the greatest similarity to lignin (26).

According to Freudenberg then, spruce lignin is a dehydrogenation polymer of coniferyl alcohol. By the action of enzymes, coniferyl alcohol is first dehydrogenated, forming highly reactive quinone methide radicals (26); see Fig. 4. Indeed, it has long been recognized (79) that quinone methides readily undergo polymerization.

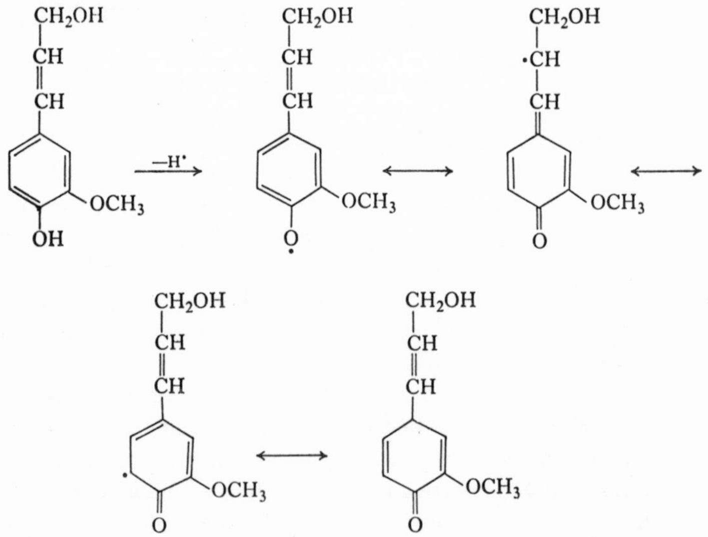

FIG. 4. Dehydrogenation of coniferyl alcohol by the action of enzymes.

These radicals could then combine in various ways, thereby forming a variety of carbon–carbon and carbon–oxygen–carbon linkages which are also believed typical for lignin.

Freudenberg also reported the isolation from his reaction mixtures, at an intermediary stage, of three "secondary lignin building stones," viz., dehydrodiconiferyl alcohol (XVI), DL-pinoresinol (XVII), and guaiacylglycerol-β-coniferyl ether (XVIII).

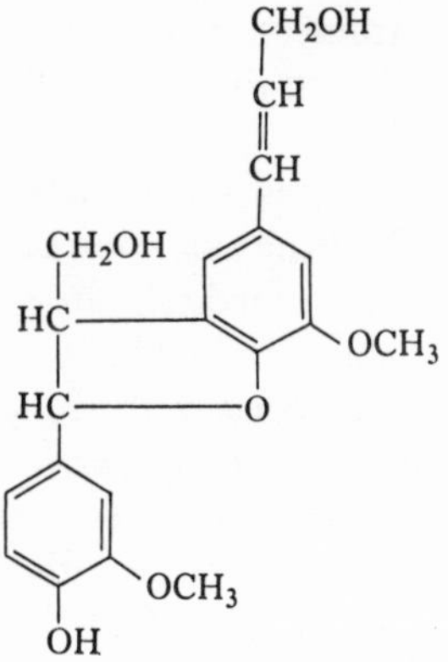

CH_2OH
CH
CH

CH_2OH
HC
HC — O — OCH_3

OCH_3
OH

(XVI)

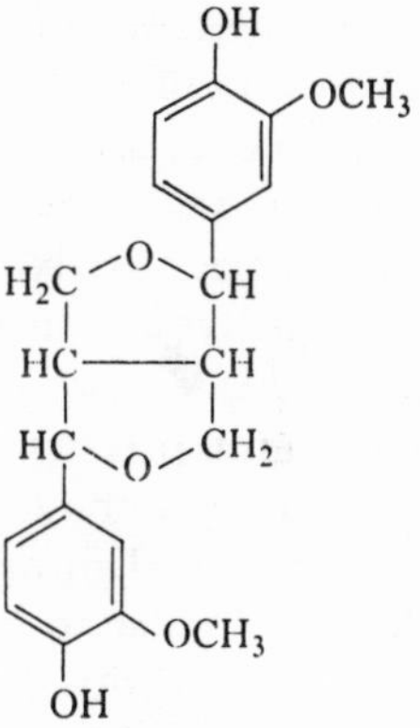

OH
OCH_3

H_2C — O — CH
HC — CH
HC — O — CH_2

OCH_3
OH

(XVII)

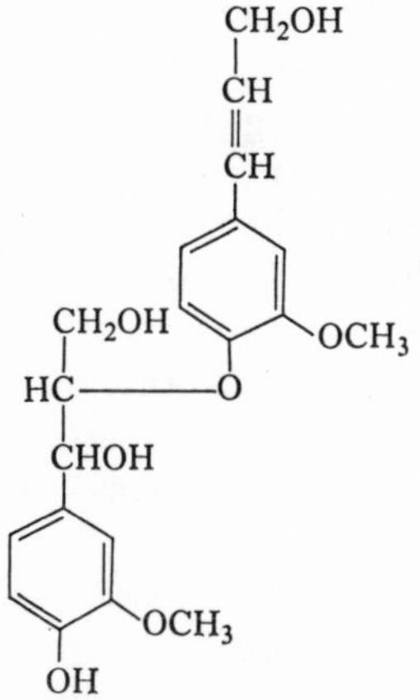

CH_2OH
CH
CH

CH_2OH
HC — O — OCH_3
$CHOH$

OCH_3
OH

(XVIII)

These secondary building stones (27) were then condensed to the dehydrogenation polymer by treatment with dehydrogenating enzymes. Furthermore, they were also reported to be present in cambial sap (28).

Accordingly, Freudenberg feels that the final phases of lignin formation are explainable (30) on the basis of an oxidation and condensation of coniferyl alcohol by certain plant enzymes.

Since the "dimeric intermediates" are also phenols, they might also be enzymically dehydrogenated to form other radicals, and these then might combine together, or with the monomeric radicals. One such product would be the trimer, guaiacylglycerol-β-pinoresinol ether. Freudenberg (31a) has claimed to have identified six such trimers. Or, after a further dehydrogenation, the dimers might combine to form tetramers, as exemplified by the formation of bisdehydropinoresinol. Freudenberg (31a) has called this process "progressive dehydrogenation."

However, many reservations and objections to this theory have been raised. For example, Nord and de Stevens found that the oxidation products of certain native and enzymically liberated lignins contained p-hydroxybenzaldehyde (73). Accordingly, they concluded that coniferyl alcohol cannot be the *only* lignin precursor, or else, that it may be preceded in the scheme of lignification by some simpler, less substituted aromatic monomer.

In addition, Erdtman (23) indicated that although Freudenberg's DHP's are similar to lignin in some respects, they are quite different in others. Aulin-Erdtman considered the "ill-defined" (4) DHP's to be stabilization products of quinone methides. She also indicated that it was not demonstrated that the DHP's were really formed by the further dehydrogenation of the dimeric dehydrogenation products of coniferyl alcohol. And, while Freudenberg claimed that these dimers were present in cambial sap, it was pointed out that the sap had already been drawn from the tree and, consequently, these compounds could actually have been nonphysiological dehydrogenation products (23) or artifacts.

In this connection, it should also be noted that numerous substances, such as eugenol, ferulic acid, and several other related compounds, which may be lignin precursors but are, at least in part, structurally different from the lignin monomers, have also been tested as "lignin progenitors" with success (72).

Also, Baylis (6) has pointed out that although Freudenberg obtained evidence for the transitory existence of quinomethide-type compounds, he did not obtain direct evidence for the existence of the postulated free radical intermediates.

Finally, Goldschmid and Hergert (36) made an examination of Western hemlock cambial constituents in an attempt to detect compounds that have been suggested as intermediates in the formation of lignin. The following compounds were identified in the cambium: quinic acid, shikimic acid, coniferin, sucrose, fructose, glucose, leucocyanidin, catechin, epicatechin, and four depsides. In addition, glucosides were detected of several α-hydroxyguaiacyl compounds. However, compounds proposed by Freudenberg as lignin intermediates, e.g., coniferyl alcohol, dehydrodiconiferyl alcohol, and guaiacylglycerol-β-coniferyl ether, were *not* found in the cambial zone.

Obviously, much still remains to be done in this area. For example, some of the earlier speculations of Freudenberg *et al.* (35) regarding the mode of adduct formation were later retracted and revised (34). Indeed, as has been noted (63a), Freudenberg's scheme of lignin formation and structure is really a hypothesis based on the status of current theory. Changes will obviously have to be made to conform with new information.

VI. THE ENZYMES INVOLVED

Parallel with the question of the chemical structures of the precursors of lignin is the biochemical problem of the nature of the enzymes which convert these precursors into lignin. The biosynthetic pathways for the formation of aromatic compounds in plants have been extensively investigated by means of tracer experiments using C^{14}-labeled compounds. It has been assumed that in plants the universally occurring aromatic amino acids are formed from carbohydrates by essentially the same pathway, i.e., Davis' scheme of aromatic biosynthesis, as in microorganisms, and that other characteristic aromatic compounds, such as lignin, are derived by secondary reactions from the aromatic amino acids, or intermediates involved in their biosynthesis. The enzymes involved in aromatic biosynthesis from nonaromatic compounds that are operative in plants and microorganisms have recently been thoroughly reviewed (41a).

As mentioned above, Freudenberg (29) observed that coniferyl alcohol may be dehydrogenated either by a mushroom or cambial sap oxidoreductase, or by peroxidase with dilute hydrogen peroxide, to a polymer which has ligninlike properties.

The oxidation of coniferyl alcohol by crude mushroom extracts has been confirmed (53), but, when a purified mushroom polyphenol oxidase (tyrosinase) was employed, the oxygen consumption was negligible. It was concluded that, although there is a heat-labile system in crude mushroom extracts which does catalyze the consumption of oxygen by coniferyl alcohol, it is not "polyphenol oxidase" that accomplishes this. Higuchi (40) suggested that the enzyme obtained from mushroom, according to the procedure used by Freudenberg, is a mixture of both laccase and tyrosinase.

Higuchi (41) investigated the properties of a phenol oxidase involved in lignification in the tissue of bamboo shoots, and found that the enzyme had a substrate specificity similar to that of laccase. He suggested that Freudenberg's enzyme may also be a laccase.

Accordingly, it is now believed (40) that the enzyme acting on coniferyl alcohol is actually a laccase. However, the known distribution of laccase in higher plants is limited. On the other hand, the system, peroxidase-hydrogen peroxide, oxidizes the same types of compounds as does laccase. Thus, Freudenberg (32) found that the yield of polymer, obtained by the action of a crude enzyme from *Araucaria excelsa* on coniferyl alcohol, could be increased severalfold by the addition of hydrogen peroxide.

Siegel (68) found that eugenol could be converted into a ligninlike product by embryonic root tips of the kidney bean in the presence of hydrogen peroxide, and he therefore attached importance to the role of peroxidase in the formation of lignin.

The chemical natures of the dehydrogenation-polymerization products of coniferyl alcohol, obtained by the action of mushroom phenol oxidase, *Rhus* laccase, and radish peroxidase, have been investigated (39). These studies indicated a close similarity among the three enzyme-formed DHP's and some resemblance to coniferous lignin. Thus, considering the wide distribution of peroxidase in higher plants, Higuchi (39, 40) suggested that peroxidase may play a more important role than laccase in lignin biosynthesis. Certain of the characteristic properties of these enzymes are reviewed in the next chapter.

VII. CONCLUSION

There can hardly be any doubt that lignin is ultimately a product of the shikimic acid pathway. The probable intervention of prephenic acid in this scheme provides an intermediate from which phenylpyruvic, *p*-hydroxyphenylpyruvic, or ferulic acid might result. These may eventually be reduced to the primary building stones such as coniferyl alcohol. An enzymic dehydrogenation-polymerization of this alcohol or of some related compound could then result in the formation of lignin.

To be sure, this analysis leaves many questions unanswered. For example, what is the source of the unmethoxylated *p*-coumaryl moieties which have been detected (62, 73)? Furthermore, hardwood lignins yield significant amounts of syringyl derivatives. What is their source ? It has been suggested (13) that the implied methoxylation of guaiacyl nuclei may occur through the intervention of methionine, but this is obviously not the only possible explanation.

REFERENCES

1. S. N. Acerbo, W. J. Schubert, and F. F. Nord, *J. Am. Chem. Soc.* **80**, 1990 (1958).
2. S. N. Acerbo, W. J. Schubert, and F. F. Nord, *J. Am. Chem. Soc.* **82**, 735 (1960).
3. E. Adler, *Tappi* **40**, 294 (1957).
4. G. Aulin-Erdtman and L. Hegbom, *Svensk Papperstid.* **59**, 363 (1956).
5. E. C. Bate-Smith, *Sci. Proc. Roy. Dublin Soc.* **27**, 165 (1956).
6. P. E. T. Baylis, *Sci. Progrs.* (*London*) **48**, 409 (1960).
7. G. Billek, *Proc. Intern. Congr. Biochem. 4th Vienna, 1958* **2**, 207 (1959).
8. F. E. Brauns, "The Chemistry of Lignin," p. 694. Academic Press, New York, 1952.
9. F. E. Brauns and D. A. Brauns, "The Chemistry of Lignin: Supplement Volume," p. 659. Academic Press, New York, 1960.
9a. S. A. Brown, *Science* **134**, 305 (1961).
10. S. A. Brown and A. C. Neish, *Can. J. Biochem. Physiol.* **33**, 948 (1955).
11. S. A. Brown, K. G. Tanner, and J. E. Stone, *Can. J. Chem.* **31**, 755 (1953).
12. S. A. Brown, D. Wright, and A. C. Neish, *Can. J. Biochem. Physiol.* **37**, 25 (1959).
13. R. V. Byerrum, J. M. Flokstra, L. J. Dewey, and C. D. Ball, *J. Biol. Chem.* **210**, 633 (1954).
14. M. Calvin, *J. Chem. Soc.* p. 1895 (1956).
15. C. J. Coscia, *Experientia* **16**, 81 (1960).
16. H. Cousin and H. Hérisséy, *Compt. Rend.* **147**, 247 (1908); *J. Pharm. Chim.* [6] **28**, 193 (1908).
17. B. D. Davis, *Advan. Enzymol.* **16**, 247 (1955).
18. B. D. Davis, *Arch. Biochem. Biophys.* **78**, 497 (1958).
19. G. Eberhardt and F. F. Nord, *Arch. Biochem. Biophys.* **55**, 578 (1955).
20. G. Eberhardt and W. J. Schubert, *J. Am. Chem. Soc.* **78**, 2835 (1956).
21. H. Erdtman, *Biochem. Z.* **258**, 172 (1933); *Ann. Chem.* **503**, 283 (1933).
22. H. Erdtman, *Svensk Papperstid.* **42**, 115 (1939); **44**, 249 (1941); *Research* (*London*) **3**, 63 (1950); H. Erdtman and C. A. Wachtmeister *Festschr. Arthur Stoll*, p. 145 (1957).
23. H. Erdtman, *Ind. Eng. Chem.* **49**, 1385 (1957).
24. H. Erdtman, *Proc. Intern. Congr. Biochem. 4th Vienna, 1958* **2**, 10 (1959).
24a. J. F. Eykman, *Chem. Ber.* **24**, 1278 (1891).
25. K. Freudenberg, *Holz Roh-Werkstoff* **11**, 267 (1953).
26. K. Freudenberg, *Fortschr. Chem. Org. Naturst.* **11**, 43 (1954).
27. K. Freudenberg, *J. Polymer Sci.* **16**, 155 (1955).
28. K. Freudenberg, *Angew. Chem.* **68**, 84 (1956).
29. K. Freudenberg, *Angew. Chem.* **68**, 508 (1956).
30. K. Freudenberg, *Ind. Eng. Chem.* **49**, 1384 (1957).
31. K. Freudenberg in *Proc. Intern. Congr. Biochem. 4th Vienna, 1958* **2**, 121 (1959).
31a. K. Freudenberg, *in* "The Formation of Wood in Forest Trees" (M. H. Zimmermann, ed.), p. 203. Academic Press, New York, 1964.
32. K. Freudenberg, *Chem. Ber.* **84**, 472 (1951); **85**, 641 (1952).
33. K. Freudenberg and F. Bittner, *Chem. Ber.* **86**, 155 (1953).
34. K. Freudenberg and M. Friedmann, *Chem. Ber.* **93**, 2138 (1960).

35. K. Freudenberg, B. Lehmann, and A. Sakakibara, *Chem. Ber.* **93**, 1354 (1960); *Ann. Chem.* **623**, 129 (1959).

36. O. Goldschmid and H. L. Hergert, *Tappi* **44**, 858 (1961).

37. C. E. Hartt and G. O. Burr, *Proc. Intern. Botan. Congr. 7th Stockholm, 1950,* p. 748 (1953).

38. M. Hasegawa, T. Nakagawa, and S. Yoshida, *Nippon Ringaku Zasshi* **39**, 159 (1957).

39. T. Higuchi, *J. Biochem. (Tokyo)* **45**, 575 (1958).

40. T. Higuchi, *Proc. Intern. Congr. Biochem. 4th Vienna, 1958* **2**, 161 (1959).

41. T. Higuchi, I. Kawamura, and H. Ishikawa, *Nippon Ringaku Zasshi* **35**, 258 (1953); T. Higuchi, T. Kawamura, and I. Morimoto, *ibid.* **37**, 446 (1955); T. Higuchi, *Plant Physiol.* **10**, 364 (1957).

41a. T. Higuchi and I. Kawamura, *in* "Moderne Methoden der Pflanzenanalyse" (K. Paech and M. V. Tracey, eds.), Vol. VII, p. 260, 1964.

42. M. Katagiri, *J. Biochem. (Tokyo)* **40**, 629 (1953).

43. M. Katagiri and R. Sato, *Science* **118**, 250 (1953).

44. P. Klason, *Svensk Kem. Tidskr.* **9**, 135 (1899).

45. P. Klason, *Chem. Ber.* **53**, 706 (1920).

46. K. Kratzl and G. Billek, *Tappi* **40**, 269 (1957).

47a. K. Kratzl and H. Faigle, *Monatsh. Chem.* **90**, 768 (1959).

47b. K. Kratzl and H. Faigle, *Z. Naturforsch.* **15b**, 4 (1960).

48. K. Kratzl and J. Zauner, *Holzforsch. Holzverwert.* **14**, 108 (1962).

49. R. E. Kremers, *Tappi* **40**, 262 (1957).

50. R. E. Kremers, *Ann. Rev. Plant Physiol.* **10**, 185 (1959).

51. S. M. Manskaja, *Dokl. Akad. Nauk S.S.S.R.* **62**, 369 (1948).

52. S. M. Manskaja, *Proc. Intern. Congr. Biochem. 4th Vienna, 1958* **2**, 215 (1959).

53. H. S. Mason and M. Cronyn, *J. Am. Chem. Soc.* **77**, 491 (1955).

54. S. Mitsuhashi and B. D. Davis, *Biochim. Biophys. Acta* **15**, 54 (1954).

55. V. Moses and M. Calvin, *Arch. Biochem. Biophys.* **78**, 598 (1958).

56. A. C. Neish, in discussion, *Proc. Intern. Congr. Biochem. 4th Vienna, 1958* **2**, 213 (1959).

57. A. C. Neish, *Ann. Rev. Plant Physiol.* **11**, 55 (1960); *in* "The Formation of Wood in Forest Trees" (M. H. Zimmermann, ed.), p. 219. Academic Press, New York, 1964.

58. F. F. Nord and W. J. Schubert, *Tappi* **40**, 285 (1957).

59. F. F. Nord and W. J. Schubert, *Proc. Intern. Congr. Biochem. 4th Vienna, 1958* **2**, 189 (1959).

60. F. F. Nord and W. J. Schubert, *Experientia* **15**, 245 (1959).

60a. F. F. Nord and W. J. Schubert, *in* "Comparative Biochemistry" (M. Florkin and H. S. Mason, eds.), Vol. IV, p. 65. Academic Press, New York, 1962.

60b. F. F. Nord and W. J. Schubert, *in* "Biogenesis of Natural Compounds" (P. Bernfeld, ed.), p. 693. Macmillan (Pergamon), New York, 1963.

61. F. F. Nord, W. J. Schubert, and S. N. Acerbo, *Naturwissenschaften* **44**, 35 (1957).

62. F. F. Nord and G. de Stevens, *in* "Handbuch der Pflanzenphysiologie" (W. Ruhland, ed.), p. 389. Springer, Berlin, 1958.

63. F. F. Nord and J. C. Vitucci, *Advan. Enzymol.* **8**, 253 (1948).

63a. I. A. Pearl, *Chem. Eng. News* **42**, No. 27, p. 81 (1964).

64. W. J. Schubert and S. N. Acerbo, *Arch. Biochem. Biophys.* **83**, 178 (1959).

65. W. J. Schubert, S. N. Acerbo, and F. F. Nord, *J. Am. Chem. Soc.* **79**, 251 (1957).

66. W. J. Schubert and F. F. Nord, *Advan. Enzymol.* **18**, 349 (1957).
67. H. Shimazono, W. J. Schubert, and F. F. Nord, *J. Am. Chem. Soc.* **80**, 1992 (1958).
68. S. M. Siegel, *Physiol. Plantarum* **6**, 134 (1953); **7**, 41 (1954); **8**, 20 (1955); *J. Am. Chem. Soc.* **78**, 1753 (1956); *Quart. Rev. Biol.* **31**, 1 (1956).
69. D. B. Sprinson, *Advan. Carbohydrate Chem.* **15**, 235 (1960).
70. P. A. Srere, J. R. Cooper, V. Klybas, and E. Racker, *Arch. Biochem. Biophys.* **59**, 535 (1955).
71. P. R. Srinivasan, M. T. Shigeura, M. Sprecher, D. B. Sprinson, and B. D. Davis, *J. Biol. Chem.* **220**, 477 (1956).
72. H. A. Stafford, *Plant Physiol.* **35**, 108, 612 (1960).
73. G. de Stevens and F. F. Nord, *Proc. Natl. Acad. Sci. U.S.* **39**, 80 (1953).
74. J. E. Stone, *Can. J. Chem.* **31**, 207 (1953).
75. J. E. Stone, M. J. Blundell, and K. G. Tanner, *Can. J. Chem.* **29**, 734 (1951).
76. F. Tiemann and B. Mendelsohn, *Chem. Ber.* **8**, 1127, 1136, 1139 (1875).
77. A. von Wacek, O. Härtel, and S. Meralla, *Holzforschung.* **7**, 58 (1953); **8**, 65 (1954).
78. A. B. Wardrop, *Tappi* **40**, 225 (1957).
79. T. Zincke and O. Hahn, *Ann. Chem.* **329**, 1 (1903).

Chapter IV • The Microbiological Degradation of Lignin

The presence of fungal decay in wood affects the appearance, physical properties, chemical composition, and microscopic structure of the wood (13, p. 53). Thus, decay in wood is usually accompanied by a change in color; the wood may become either bleached or darkened. Decayed wood is softer and less strong than sound wood; in an advanced stage of decay, it crumbles easily. Decayed wood is less dense than sound wood, due to the partial destruction of the substance of the wood itself by the metabolic activities of the infecting fungus; thus, wood in an advanced state of decay may be extremely light in weight, but retain its original outward form and structure. Decayed wood absorbs water and becomes waterlogged much more readily than sound wood (13).

Decomposition of wood by fungi is of two main types, which have been described as "brown rots" and "white rots" (13). In brown rot, the cellulose and other associated carbohydrates are attacked preferentially; the lignin remains in a relatively unchanged condition, and the decaying residue gradually turns brown in color. In white rot, all components of the wood, including the lignin, are decomposed; sometimes in the residue, patches of a white substance may be observed; this material has been referred to as an almost pure form of cellulose. Classification of the types of wood decay on this basis was suggested as early as 1927 (22, 76) and the terms "destruction rot" and "corrosion rot" were also suggested for brown rot and white rot, respectively.

While various individual species of wood-rotting fungi may differ from one another in the extent to which they penetrate and degrade the cell walls of wood, their action, within the two groups of the brown and white rots, appears to be generally similar. Thus, in wood attacked by a fungus which causes brown rot, the cell wall does not appear to be thinned appreciably until a very late stage of decay; and, even after almost all the cellulose has been removed, the residual lignin preserves the form of the original cell wall (13, p. 62).

In white rots, on the other hand, a general thinning of the cell walls is usual even in the early stages of decay. This is to be anticipated, since all constituents of the wood are decomposed by white-rot fungi. However, the residual cellulose left in wood attacked by white rots is sufficient, even in an advanced stage of decay, to enable the wood to retain its shape and outward structure; whereas in brown rot, disintegration due to the removal of cellulose begins to appear at an earlier stage of decay (13, p. 62).

Thus, the general effects of the two types of decay, the brown and the white rots, are now fairly well known, and the overall results of the process of wood decay can be described; but at this time, much less information is available about the chemistry of the intermediate stages occurring during the fungal decay of wood. This aspect of the phenomenon will be considered in some detail in subsequent sections of this chapter.

I. LIGNIN-DEGRADING FUNGI

The fungi belong to a class of the plant kingdom which differs from other plants in that they do not produce chlorophyll, and, hence, are unable to synthesize their own organic nutrients photosynthetically. Instead, they must derive their nourishment from previously existing organic molecules; hence, wood-destroying fungi are found in nature growing either as parasites on other living plants, such as trees, or else as saprophytes on the nonliving remains of plants, such as dead wood.

Fungi reproduce as a result of the germination of spores, which are produced in a variety of ways, and they are classified according to how these spores are borne. Most of the fungi which cause the decay of wood belong to one of the groups of "higher fungi" or Basidiomycetes, in which the spores are borne on small club-shaped structures, or basidia. The subdivision of the Basidiomycetes in turn is based on the shape and form taken by the hymenium, which is a compact superficial layer on which the basidia are formed. Most wood-destroying fungi belong to one of four families of the Basidiomycetes, namely, the Thelephoraceae, Hydnaceae, Polyporaceae, and Agaricaceae. In addition to the Basidiomycetes, a few species of fungi which are also of some importance as causes of wood rot appear in the other principal class of higher fungi, namely, the Ascomycetes (13, p. 9).

Approximately 2000 individual species of Basidiomycetes are known to cause wood rot. These may grow either on living or dead trees, and may destroy either the total wood as a whole, or by more specific action, individual constituents of it. Extensive tabulations of the species of Basidiomycetes which attack wood and cause rots of both the brown and the white types have appeared (13, 49). Other causative agents for the biological decomposition of wood components are the "digestive" microorganisms which inhabit the digestive tracts of certain animals, and microorganisms of soil and water. These will not be reviewed here.

II. ENZYMES OF LIGNIN-DEGRADING FUNGI

The enzymes of wood-destroying fungi have received relatively little systematic investigation in comparison with those of many other classes of organisms, and most of the information about them is of a rather general nature. Relatively little is known about their chemistry or modes of activity. Most of the early investigators confined themselves to testing extracts of the fungi on various substrates, and if chemical changes in these substrates were observed, the existence of a specific enzyme was assumed and an appropriate name was devised for it. For instance, if the lignin of wood was found to be altered, the presence of a specific "lignase" enzyme was presumed (13, p. 30). (But the existence of one specific lignin-degrading enzyme is highly unlikely. The nature of the enzymes believed to be responsible for its degradation will be considered below).

Zeller (78) was among the first to investigate the enzymes of a wood-rotting fungus. He recorded the following list of enzymes in *Lenzites sepiaria*: esterase, maltase, invertase, raffinase, emulsin, tannase, diastase, cellulase, hemicellulase, pectinase, inulinase, urease, hippuricase, nuclease, protease, erepsin, trypsin, rennet, catalase, tyrosinase, and "oxidase." With the exception of the last three, he found greater amounts of these enzymes present in the vegetative mycelium than in the sporophore tissue.

A comparison of the intracellular and extracellular enzyme activity in eight species of Polyporaceae at different stages of growth (10) revealed that the relative amount of extracellular enzymic activity was invariably greater than that of the corresponding intracellular activity, indicating that the enzyme was secreted into its surrounding medium. And, of course, this is to be expected, since the high molecular weight, water-insoluble wood constituents, such as lignin and cellulose, can be utilized by the organism only after they have been rendered water soluble (13, p. 30).

Certain of the extracellular enzymes of white-rot fungi give rise to a well-known, characteristic color test, known as the "Bavendamm reaction" (6). In this reaction, a white-rot fungus, when cultivated on an agar medium containing gallic and tannic acids, or certain other phenolic substances, produces a dark-colored zone around the mycelium; a brown-rot fungus does not form this colored zone. This phenomenon has been extensively studied in an effort to identify the responsible enzyme(s).

Thus, in agreement with Bavendamm, it was observed that lignin-decomposing fungi, in the presence of tannins, excrete enzymes of a "phenoloxidase" nature (16), whereas other wood-destroying fungi, such as *Merulius lacrymans*, which decompose the cellulose of wood but not the lignin, do not produce such enzymes. Further examination (53) revealed that the lignin-decomposing fungi excrete phenoloxidases into the medium, whereas *mycorrhizal* molds, with few exceptions, do not. These "phenoloxidases" were believed to consist of at least two principal enzymes, namely, tyrosinase and laccase (53). (More recent studies on the enzymology of the Bavendamm reaction will be discussed below.) In the interim, enzymes identified as laccase (48), tyrosinase (79), or simply as phenoloxidase (24) have been reported to be produced by a wide variety of wood-destroying fungi of the white-rot type.

The production of peroxidase by lignin-decomposing fungi has also been reported (58); thus, this enzyme was found in fourteen of one hundred and eighty-two species examined, while tyrosinase was present in sixty-two of them. The peroxidase produced by the fungi was reported to be similar to that present in horse radish.

Obviously, much research must be undertaken on the enzymes of the wood-rotting fungi, both in regard to the natures of the enzymes themselves, and the chemistry of their activity. However, unquestionably, the two enzymes that have most frequently been implicated in the degradation of lignin by fungi are tyrosinase (or, as it will hereinafter be called, phenolase) and laccase. Both phenolase and laccase are copper-containing proteins.

The copper proteins, in general, comprise many enzymes with an "oxidase" function, and include phenolase, laccase, and ascorbic acid oxidase, all of which are of plant origin, and also several proteins obtained from animal tissues or fluids, whose functions are uncertain (17). The copper-containing enzymes have the common property of catalyzing the direct oxidation of their respective substrates by atmospheric oxygen. They do not function anaerobically, i.e., they will not bring about the oxidation of their substrates via methylene blue or similar dyes, and their action does not result in the formation of hydrogen peroxide. They are sensitive to cyanides. Hence, they are classified as oxidases, rather than as dehydrogenases (17).

A. Phenolase

The enzyme phenolase (44) has in the past been referred to as tyrosinase (17), monophenoloxidase (71), polyphenoloxidase (71), phenolase complex (60), phenolase system, catechol oxidase, and dopa oxidase (44). This enzyme was discovered and named "tyrosinase" in 1895 (9), when it was demonstrated that the darkening of a variety of mushroom, *Russula nigricans*, was due to an enzymic oxidation of the phenolic amino acid, tyrosine. The enzyme has since been identified in bacteria, molds, higher plants, crustaceans and mollusks. Good plant sources of the enzyme are wheat bran, the white potato, and certain mushrooms, including the common variety, *Psalliota campestris* (17).

Phenolase is characterized by its ability to catalyze the aerobic oxidation

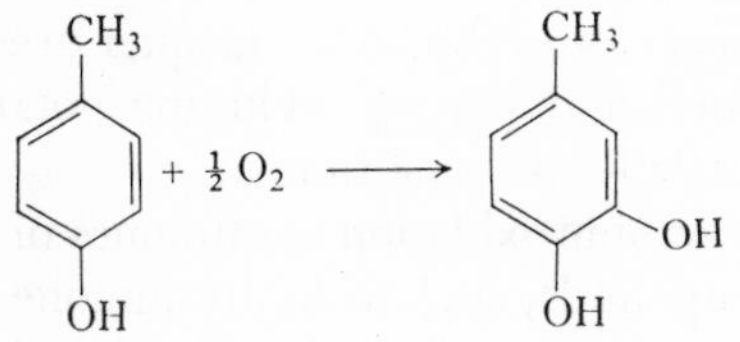

$$(1)$$

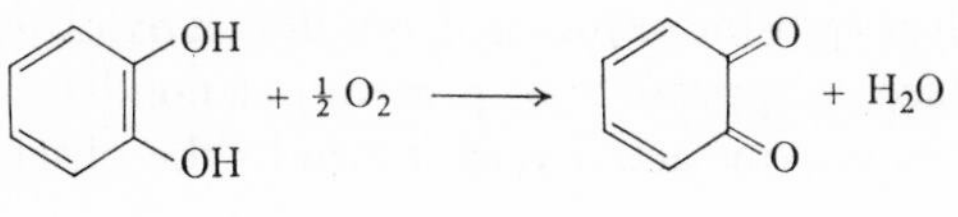

$$(2)$$

of both monohydric and o-dihydric phenols. Further, it is generally conceded that this enzyme can catalyze two fundamentally different kinds of reactions: (a) the hydroxylation of certain monohydric phenols to o-dihydric phenols [Eq. (1)] and (b) the oxidation of o-dihydric phenols to o-quinones [Eq. (2)].

These two activities are commonly referred to as the "cresolase" (i.e., monophenolase) and the "catecholase" (i.e., o-dihydric phenolase) activities of the enzyme, respectively. Thus, monophenols such as phenol itself, p-cresol and tyrosine, and the corresponding o-dihydric phenols, catechol, homocatechol, and 3,4-dihydroxyphenylalanine, are all rapidly oxidized in aqueous aerobic systems containing the enzyme.

Although it would seem likely that both activities of phenolase are probably dependent on the copper present, little is known about the way in which the copper is bound within the enzyme, or its role in the monophenolase activity (17). In the case of the o-dihydric phenolase activity, as the result of inhibition studies, it was postulated (47) that during the course of the reaction, the copper of the enzyme is first reduced to the cuprous state, and is later reoxidized by molecular oxygen.

B. Laccase

In 1883, it was observed (77) that the latex of the Japanese lacquer trees, *Rhus succedanea* and *Rhus vernicifera*, contained a thermolabile substance that caused the darkening and hardening of the latex in air. Later, Bertrand (7) obtained from the latex of the Indo-Chinese lacquer tree an enzyme which he called "laccase." He demonstrated that the darkening of the yellow sap into the black "lac" consisted of an enzymic oxidation that involved the action of laccase on certain catechol derivatives called urushiol, hydrourushiol, and lactol.

Purified preparations of laccase have now been obtained from the latex of Indo-Chinese, Japanese, and Burmese lacquer trees. The enzyme is also found in many fungi and plants, including potatoes, beets, apples, cabbages, and several varieties of mushrooms.

Laccase catalyzes the aerobic oxidation to quinones of various polyhydric phenols and related compounds, such as hydroquinone [Eq. (3)], guaiacol, catechol, pyrogallol, and p-phenylene diamine, but it does not oxidize monophenols such as tyrosine or p-cresol, nor does it oxidize resorcinol (17).

The copper of laccase appears to be present in a nondialyzable form, and to be essential to the enzyme activity, as shown by the observation that the activities of different preparations were proportional to their copper contents (43). The activity of the purified enzyme is strongly inhibited by cyanide, sulfide, and azide, but carbon monoxide has no effect (43).

$$\text{(benzene-1,4-diol)} \quad + \; \tfrac{1}{2} O_2 \longrightarrow \text{(1,4-benzoquinone)} \quad + \; H_2O \qquad (3)$$

C. Comparison of Phenolase and Laccase Activities

In its natural sources, phenolase is often found accompanied by catalase, peroxidase, and other phenoloxidases, including laccase. Thus, Bertrand (8) observed that laccase and tyrosinase were frequently found together in many plant tissues, and he used the colorations developed during the oxidation of tyrosine or hydroquinone as a means of distinguishing the two enzymes. (Phenolase has no action on hydroquinone.) He also reported that the two activities could be separated because of the greater stability of laccase toward heat and alcohol.

Keilin and Mann (43) compared the properties of a purified laccase preparation with those of a "polyphenoloxidase" which they obtained in a purified form from cultivated mushroom. Several differences were apparent. The laccase solutions were blue in color, while polyphenoloxidase solutions of comparable concentration and purity were yellow. The laccase oxidized p-phenylene diamine more readily than it did catechol, whereas polyphenoloxidase had little or no action on p-phenylene diamine and showed its maximum activity on catechol. The laccase oxidized hydroquinone but had no action on the monophenols, p-cresol and tyrosine. In contrast to this, the polyphenoloxidase did not oxidize hydroquinone, and slowly oxidized monophenols after a lag period (17).

Furthermore, although carbon monoxide strongly inhibited polyphenoloxidase activity, it had no effect on laccase. Thus, it might appear that the activity of laccase does not depend on a reversible reduction and oxidation of its copper, as has been proposed for polyphenoloxidase activity (17, 47), but, more recently, a change in the valence state of the copper in laccase, occurring during its enzymic activity, has been reported (59).

Thus, in summary, phenolase may be distinguished from laccase and other oxidases through its ability to catalyze two essentially different oxidations, namely, the introduction of a hydroxyl group into monohydric phenols *ortho* to the one already present [Eq. (4)], and the oxidation of *o*-dihydric phenols to the corresponding *o*-quinones [Eq. (5)]. Laccase, on the other hand, can directly accomplish the oxidation to quinones of *o*-dihydric phenols [Eq. (5)], *p*-dihydric phenols [Eq. (6)], and polyhydric phenols [Eq. (7)], but it is without effect on monohydric phenols.

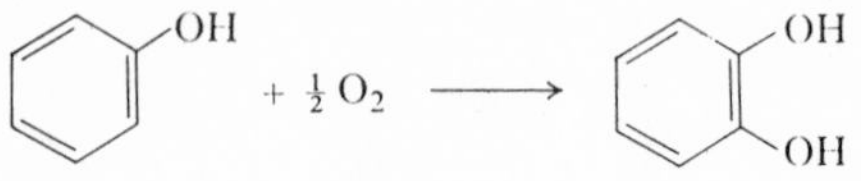

$$\text{(structure)} + \tfrac{1}{2}O_2 \longrightarrow \text{(structure)} \qquad (4)$$

$$\text{(structure)} + \tfrac{1}{2}O_2 \longrightarrow \text{(structure)} + H_2O \qquad (5)$$

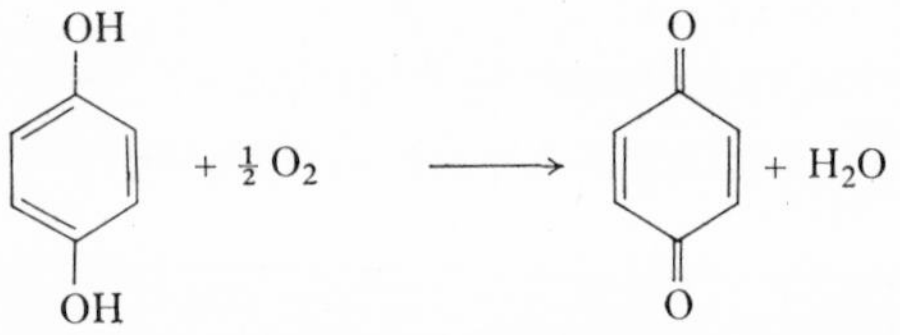

$$\text{(structure)} + \tfrac{1}{2}O_2 \longrightarrow \text{(structure)} + H_2O \qquad (6)$$

$$\text{(structure)} + \tfrac{1}{2}O_2 \longrightarrow \text{(structure)} + H_2O \qquad (7)$$

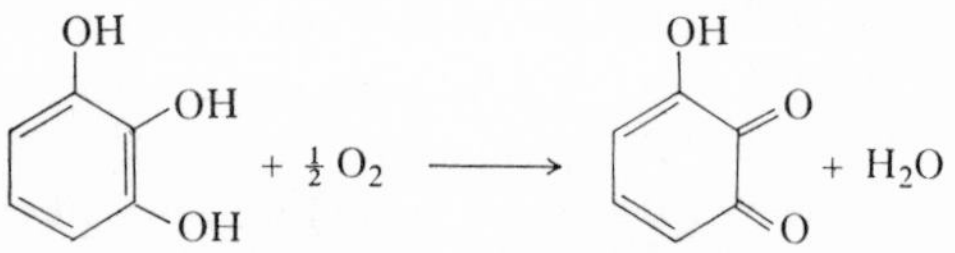

D. Occurrence of Phenolase and Laccase in Lignin-Degrading Fungi

Davidson *et al.* (16) found that 96 % of the white-rot fungi investigated by them gave a positive Bavendamm reaction, and they concluded that laccase catalyzes the oxidation of lignin. However, since laccase formation by fungi appears to be dependent on the chemical composition of the substrate, these results have not necessarily been considered as conclusive. The failure of certain fungi containing laccase to decompose lignin may be due to their lack of other enzymes necessary for its decomposition.

In an investigation of the occurrence of phenolase and laccase in a number of Hymenomycetes, it was found (55) that the fruiting bodies of some of them formed both enzymes. The mushroom, *Psalliota bispora*, was found (54) to contain different phenoloxidases in its various parts; the fruiting bodies contained phenolase, while mycelia cultivated in nutrient solution contained only laccase; in the rhizomorphs, both enzymes were present.

Dion (19) investigated the filtrates of *Polyporus versicolor* and found that they gave the reactions of laccase, but probably contained other enzymes as well. According to Fåhraeus (21), the phenoloxidase produced by lignin-decomposing fungi is identical with laccase.

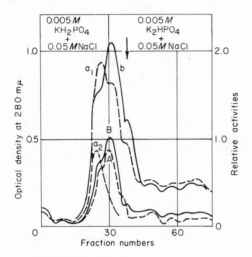

FIG. 1. Column chromatography of the crude enzyme preparations isolated from the mycelial pellets of *Fomes fomentarius* and from its mycelia-free culture filtrate. Dashed line, mycelial pellets; solid line, culture filtrate. *A* and *B*, Optical density of effluent; a_1 and b, relative activity of laccase in effluent; a_2, relative activity of peroxidase in effluent. Column dimensions: 2 cm. diameter, 50 cm. length. Effluent collected: 5–5.5 ml. per tube. Activities were measured by adding 3 ml. of sample to 2 ml. of substrate and following the extinction change at 415 mμ or 430 mμ.

Fungal oxidase enzymes have recently been obtained from acetone powders prepared from both the mycelial pellets and culture filtrates of *Fomes fomentarius* and *Collybia velutipes* (42). The principal enzymes obtained from *F. fomentarius* included phenolase, a laccase-like enzyme, and peroxidase, while only traces of the first two were detected in *C. velutipes*. Phenolase activity in all preparations was small (Figs. 1 and 2).

In general, more of the laccase-like enzyme was present in the culture filtrates than in the mycelia. Paper electrophoresis indicated the presence of

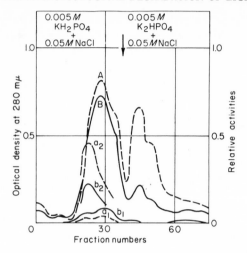

FIG. 2. Column chromatography of the crude enzyme preparations isolated from the mycelial pellets of *Collybia velutipes* and from its mycelia-free culture filtrate. Dashed line, mycelial pellets; solid line, culture filtrate. *A* and *B*, Optical density of effluent; a_1 and b_1, relative activity of laccase in effluent; a_2 and b_2, relative activity of peroxidase in effluent.

three or four laccase-like enzymes in protein fractions obtained from both the pellets and filtrates of *F. fomentarius* (Fig. 3). The purified laccase-like enzyme solution had a maximum absorption at 280 mμ and was light yellow in color. The enzyme oxidized hydroquinone and guaiacyl compounds, was

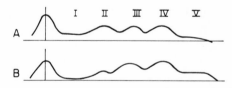

FIG. 3. Paper electrophoresis of enzyme fractions of *Fomes fomentarius*, purified chromatographically. *A*, Enzyme fraction isolated from mycelial pellets; *B*, enzyme fraction isolated from culture filtrate. *I*, Tyrosinase; *II–V*, laccase-like enzymes.

not inactivated by carbon monoxide, and did not oxidize *p*-cresol or tyrosine under aerobic conditions. The presence of copper in the enzyme was established (42). Purified enzyme preparations, isolated from the mycelia and filtrates of both fungi, exhibited an absorption peak in the region of 410–440 mμ on reaction with dithionate, indicating the presence of peroxidase.

The Bavendamm reaction. Higuchi (34) made a thorough investigation of the enzymology involved in the Bavendamm reaction. He found that although a preparation of tyrosinase isolated from potatoes did not oxidize hydroquinone or resorcinol, a purified enzyme preparation isolated from the mycelium of *Coriolus hirsutus* (the activity of which was not inhibited by carbon monoxide) did oxidize polyphenols such as catechol, hydroquinone, eugenol, and guaiacol, but not tyrosine. This enzyme was therefore believed to be a laccase. But the activity of the *crude* enzyme of the organism *was* partially inhibited by carbon monoxide, and, therefore, probably contained some phenolase.

Higuchi (38) also tested the filtrates, mycelial extracts, and mycelial residues of twenty white-rot and nine brown-rot fungi for their laccase and phenolase contents by examining their oxidizing activities on tyrosine, catechol, guaiacol, *p*-cresol, hydroquinone, α-naphthol, and gallic and tannic acids. The filtrates of certain of both the brown-rot and the white-rot fungi oxidized tyrosine. Although the white-rot fungi, in media containing α-naphthol (for the laccase test) and tyrosine or *p*-cresol (for the phenolase test), gave positive reactions for both enzymes, the brown-rot fungi gave no laccase reaction, although some species gave a positive phenolase reaction.

The Bavendamm reaction, therefore, seems to be caused by the presence of both laccase and phenolase, but the amount of the latter that is required is small when compared with that of the former. Tannic and gallic acids are oxidized by both laccase and phenolase, and cannot be used for a differentiation between the two enzymes; α-naphthol was therefore suggested for the laccase test and *p*-cresol for the phenolase test. The classic "Bavendamm reaction" can therefore be divided into individual laccase and phenolase reactions, and can be used to test for these enzymes (35).

The ability to impart a color to tannic acid has more recently been used by Henderson (31) as a criterion in the isolation of lignin-decomposing fungi from soil samples. But no correlation was found between the presence in the fungus of enzymes which oxidize tannic acid and the ability of the organism to attack lower molecular weight "lignin-related" compounds. This is understandable since the oxidation of phenolic compounds ordinarily involves quinone formation and polymerization (and resultant pigment formation), while the attack on the phenolic compounds studied by Henderson (32) involves the rupture of a benzenoid ring, obviously two quite different processes. Therefore, it would appear that the use of the tannic acid method in the isolation of lignin-decomposing fungi has little foundation. Indeed, when lignin is degraded by white-rot fungi, the remaining decayed material is lighter in color than the original wood (33).

III. CHEMISTRY OF THE ENZYMIC DEGRADATION OF LIGNIN

Because of the "bleaching" action of some of the white-rot fungi on wood, it had long been assumed that they mainly attacked the lignin of wood, but it has since been shown that while the fungi which cause white rot do decompose lignin, other substances are simultaneously attacked, and no wood-rotting fungus is known to derive all of its energy from lignin alone. Undeed, it is not likely that such a fungus will ever be found, since lignin is a comparatively stable and inert material which is consequently relatively resistant to enzymic decomposition (13, p. 60).

However, since the chemistry and enzymology of the microbiological assimilation of the carbohydrate components of wood have been thoroughly treated (26, 63, 68), the following will be devoted to a consideration of the disintegration of the lignin component of wood only.

A. Over-All Effects of Lignin Decay

It has been known for some time (57) that as the decay of wood by white-rotting fungi proceeded, the characteristic phloroglucinol-hydrochloric acid color reaction exhibited by the wood gradually diminished. Thus, it appeared that the chemical entity responsible for this reaction was affected by the fungal activity. Since it is now believed that the responsible group is an aldehyde, and since the degradation of lignin involves enzymes of an oxidase nature, it seems reasonable to conclude that at least one aspect of lignin disintegration involves the oxidation of an aldehyde to a carboxylic acid. Furthermore, it has long been known that the lignin in decaying wood is rendered more soluble in dilute aqueous alkali as the fungal decay proceeds (63), implying the formation of acid groups.

One of the earliest investigations of the metabolic products of wood-destroying fungi was carried out by Czapek (14), who found that increased amounts of "hadromal" could be extracted from sprucewood after it had been attacked by fungi. "Hadromal" is now believed to correspond to coniferyl aldehyde (11, p. 36).

During its fungal degradation, lignin appears to undergo both oxidation and demethoxylation. Indeed, in many cases, the extent of the disintegration of lignin has been estimated by the decrease in its methoxyl content. It had been claimed (75) that in the aerobic microbial decomposition of lignified plant materials, the methoxyl content of the residual lignin was not modified to any appreciable extent, whereas under anaerobic conditions, the methoxyl content was gradually reduced. However, in experiments with isolated lignins, from 23 to 41% of the methoxyl content of the lignins was lost after 6 months of aerobic decay (5).

The lignin of sprucewood which had been decayed by various fungi was studied by Apenitis *et al.* (4). They found that the alkaline nitrobenzene oxidation of certain extracts of the decayed wood gave higher yields of vanillin than did the oxidation of the unextractable portion of the lignin itself. These observations were interpreted on the basis of the results of the analogous oxidation of certain synthetic "lignin-like model compounds" of the type (I).

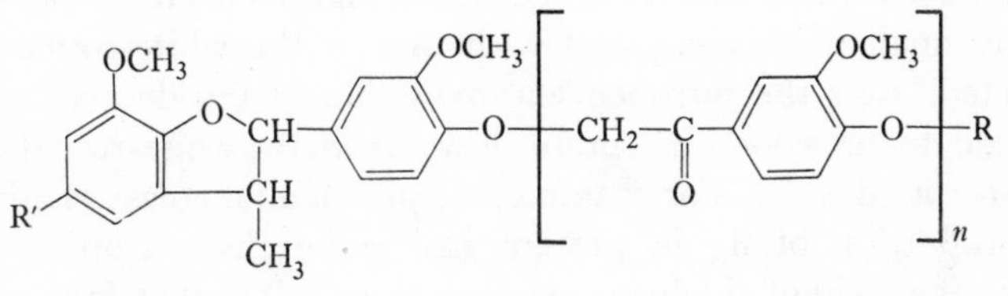

(I)

The structures of these compounds could be varied, and products containing moieties either of the "open" (II) type or of both "open" and "condensed" (III) types were prepared (50). These model compounds were

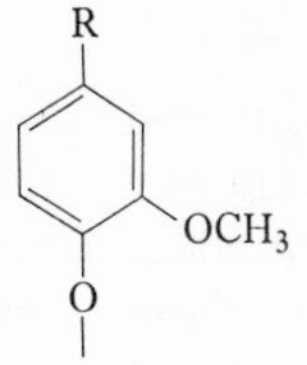

(II) 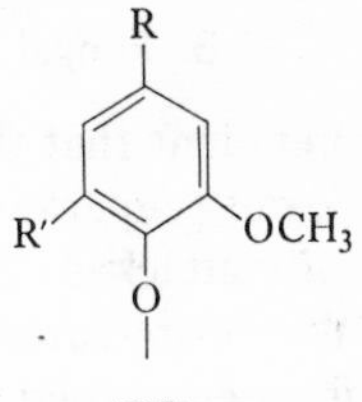   (III)

then subjected to oxidation with nitrobenzene and alkali, and the yields of vanillin were determined (51). From these studies, it appeared that no appreciable splitting of C—aryl bonds occurred during the oxidation, but that other C—C linkages were split fairly readily. Consequently, structures of type (III), containing an "extra" C—aryl bond, gave only insignificant yields of vanillin.

The lower yield of vanillin from the decayed wood lignin was then interpreted by Leopold (52) as due to the fungal attack taking place preferentially on the "open" elements of lignin; thus, decayed wood lignin becomes depleted in vanillin-yielding moieties, and concomitantly, enriched in nonvanillin yielding moieties.

Similarly, Enkvist *et al.* (20) found that the oxidation of fungally decayed wood with nitrobenzene and alkali gave only 4.2% vanillin, as compared with 6.4% from sound wood of the same species, thereby indicating that a proportional amount of vanillin-yielding groups of the lignin had been destroyed during the fungal decay.

Higuchi and Kawamura (36) compared the lignin residues in beechwood samples which had been exposed to attack by various brown- and white-rot fungi. The results revealed that the lignin was decomposed substantially by the white-rot fungi, but only slightly, or not at all, by the brown-rots, and that the loss in methoxyl content in the lignins from the brown-rotted wood was small when compared with that of the white-rotted material. As expected, after the nitrobenzene oxidation of the decayed wood, the yields of aldehydes were considerably lower from the white-rotted than from the brown-rotted wood. For example, in spite of an increase of almost 50% in the proportion of lignin present in the beechwood attacked by *M. lacrymans*, the yield of aldehydes was less than half of that from the sound wood. Such observations induced Higuchi *et al.* (37) to conclude, in agreement with Leopold (50–52) that the "open" vanillin-yielding moieties of lignin are preferentially attacked by fungi.

B. Fungal Degradation of Isolated Lignins

It is quite evident that the lignin in wood is destroyed by many wood-rotting fungi, but preparations of lignin isolated from wood are, in general, not very readily attacked. Nevertheless, many attempts have been made to cultivate white-rotting fungi on chemically defined media containing isolated lignin preparations as their sole source of organic material.

After screening a large number of white-rotting fungi by cultivating them in a medium in which lignin was the limiting source of carbon, Day *et al.* (18) found that *Polyporus abietinus* and *Poria subacida* grew satisfactorily. They then grew these organisms on a medium in which lignin was the only source of carbon by a process of adaptation in which the fungi were grown on media containing mixtures of lignin and glucose, with the amount of the latter being progressively decreased until the fungi were growing on lignin alone (28). Later they claimed (64) that *P. versicolor*, without previous adaptation, could be grown in a nutrient medium in which lignin was the only source of carbon. They also reported that *Polyporus abietinus* and *Poria subacida* could become adapted to some of sixteen lignin preparations isolated by various laboratory and commercial procedures (29).

In similar experiments, Dion (19) found that cultures of *P. versicolor* could be grown on isolated native aspen lignin as the sole source of carbon. Van Vliet (74) also found that *P. versicolor* could be cultivated on a medium employing Brauns' lignin as its sole organic material.

In experiments on the degradation of softwood lignins by white-rot fungi (39), three such organisms, *F. fomentarius*, *P. subacida* and *Trametes pini*, were cultivated on a basal medium supplemented with one of the following lignin preparations: pine native lignin, spruce native lignin, pine milled-wood lignin, and spruce milled-wood lignin.

In subsequent studies, several other white-rot fungi were incubated in media containing gradually decreasing amounts of glucose and increasing amounts of pine lignin as their source of carbon. All the fungi tested were adaptable to lignin utilization by growing them in the lignin-containing medium for a prolonged period. "Polyphenoloxidase-rich" organisms grew faster than "polyphenoloxidase-poor" species (40). On incubating *P. versicolor*, *P. hirsutus*, and *P. subacida* J247 in a medium containing pine native or pine milled-wood lignin, the amount of residual lignin decreased from 68 to 45% of the original, whereas the degree of utilization by *P. subacida* N199, *F. fomentarius*, *Fomes annosus*, and *T. pini* varied between 31 and 18% (40).

Analyses of the pine native and milled-wood lignins after decay by *P. subacida* N199, *F. fomentarius*, *F. annosus*, and *T. pini* revealed that the methoxyl contents of these lignins were higher than were those of the lignins decayed by *P. versicolor*, *P. hirsutus*, and *P. subacida* J247. Furthermore, the yields of vanillin obtained after nitrobenzene oxidation of the pine native and milled-wood lignins degraded by the first four fungi were considerably greater than from the lignins decayed by the latter three (40).

The elementary composition and constituent groups present in the pine and spruce lignins decayed by *F. fomentarius* and *P. versicolor* were determined, and it was evident that the degradation of the lignins included a loss of methoxyl groups concomitant with the formation of phenolic hydroxyls. Further, the contents of oxygen, total and phenolic hydroxyl, and carbonyl and carboxyl groups increased during the fungal decomposition. The decayed lignins also became depleted in vanillin-yielding moieties (40).

It was also observed that the decomposition of the lignins by these fungi was more accelerated with mechanical agitation than without it. In addition, the amounts of the pine and spruce milled-wood lignins that decomposed were less than the corresponding amounts of the native lignins (40).

Infrared absorption spectra of the decayed lignins revealed the presence of a large proportion of carbonyl and carboxyl groups. The carbonyl group in the decayed lignins was believed more likely to be a ketone than an aldehyde because these lignins produced only a relatively weak violet color with the phloroglucinol-hydrochloric acid reagent (40).

Fomes fomentarius and *C. velutipes* were also adapted to spruce native lignin utilization by cultivating them in the lignin-containing medium for a prolonged period. Analysis of this lignin before and after decay revealed that the methoxyl content decreased as a result of the decay, while the relative amounts of phenolic hydroxyl groups and of oxygen increased. The decayed lignins also became depleted in vanillin-yielding moieties (42).

C. Intermediate Products of Lignin Decay

As mentioned above, although the over-all effects of the two types of wood decay, the brown and white rots, are now fairly well clarified, and certain generalities regarding the phenomenon of the fungal decay of wood can be deduced, nevertheless little tangible information has been established regarding the chemistry of the intermediary stages of the fungal decay of lignin, including structures and modes of formation of the transitory intermediate products formed. Thus, an elucidation of the intermediate enzymic reactions by which lignin is degraded by fungi remains one of the goals of the investigation of the biochemistry of this substance. At the same time, intensive research on the chemistry of this material continues, and it is obvious that when the structure of lignin is better understood, it may then become possible to formulate reaction schemes for the complete gradual degradation of the complex polymer to its ultimate disintegration products.

Nevertheless, a limited number of intermediate products of the fungal degradation of lignin have been identified. Thus, Henderson (30) demonstrated the presence of vanillic (IV) and syringic acids (V) in birchwood that was decayed by *P. versicolor* and *T. pini*. Higuchi *et al.* (36, 37) observed that coniferaldehyde (VI), vanillin (VII), and syringaldehyde (VIII) were formed from beechwood lignin by several white-rot fungi. Fukuzumi (25) identified 3-methoxy-4-hydroxyphenylpyruvic acid (IX) in a spruce native lignin-containing medium decomposed by *P. subacida*, and guaiacylglycerol-β-coniferyl ether (X) and a "vanillic acid-like" compound in sprucewood meal decayed by the same fungus. Although other investigations have been reported on the biochemical degradation of lignin, and on the enzymic reactions involved (11, p. 599; 12, p. 577; 49), little integration of the information into a continuous reaction sequence has been possible.

Accordingly, in Nord's laboratory, a study of certain of the chemical properties of isolated pine and spruce lignins before and after decay by several white-rot fungi and an investigation of the intermediate products formed from these lignins by the action of the fungi were carried out. In preliminary investigations (39), three such fungi, *F. fomentarius*, *P. subacida*,

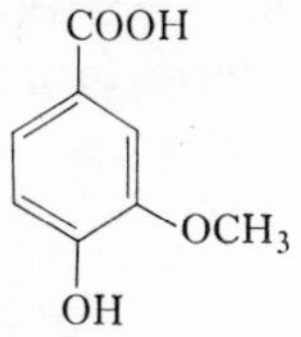

COOH

OCH₃

OH

(IV)

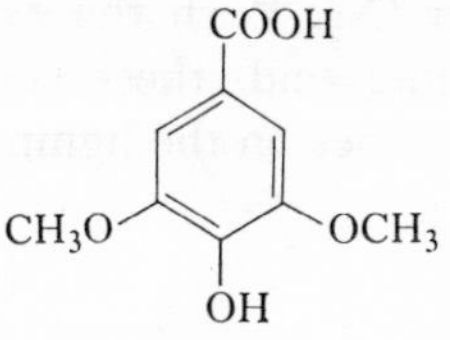

COOH

CH₃O   OCH₃

OH

(V)

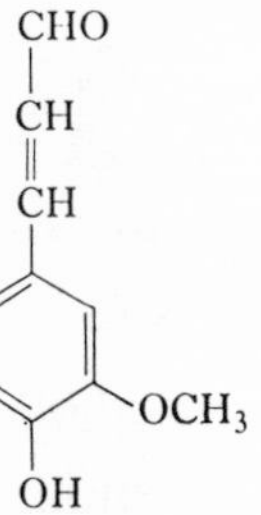

CHO

CH
‖
CH

OCH₃

OH

(VI)

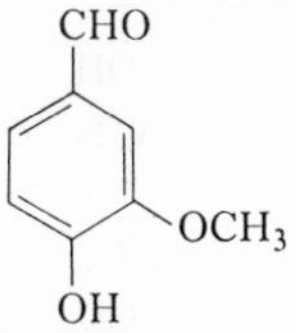

CHO

OCH₃

OH

(VII)

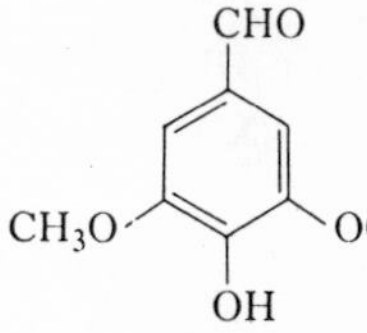

CHO

CH₃O   OCH₃

OH

(VIII)

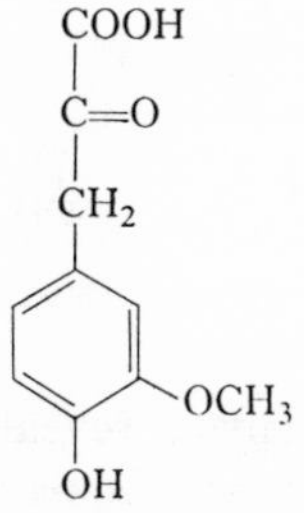

COOH
|
C=O
|
CH₂

OCH₃

OH

(IX)

CH₂OH
|
HC—O   C=C—CH₂OH

HC—OH   OCH₃

OCH₃

OH

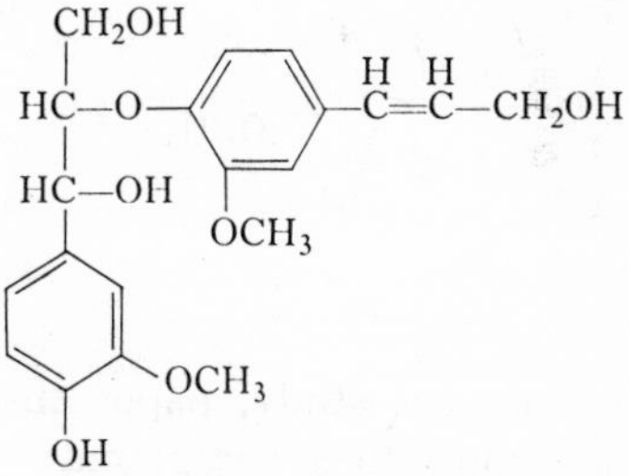

(X)

and *T. pini*, were cultivated in media containing either pine native lignin, spruce native lignin, pine milled-wood lignin, or spruce milled-wood lignin. After 28 days of growth, the resulting culture filtrates were extracted with ether, and the extracts were divided into acidic, carbonyl, phenolic, and neutral fractions. Paper chromatographic analysis revealed the presence of the following substances after decomposition by *F. fomen-tarius:* in the acidic fraction, vanillic acid (IV), *p*-hydroxybenzoic acid (XI), ferulic acid (XII), and *p*-hydroxycinnamic acid (XIII); in the carbonyl fraction, 3-methoxy-4-hydroxyphenylpyruvic acid (IX), vanillin (VII), dehydrodivanillin (XIV), coniferaldehyde (VI), and *p*-hydroxycinnam-aldehyde (XV); in the phenolic fraction, guaiacylglycerol (XVI) and its *β*-coniferyl ether (X). With the exception of *p*-hydroxybenzoic acid and *p*-hydroxycinnamic acid, these compounds were formed by the action of the fungal enzymes on the lignin.

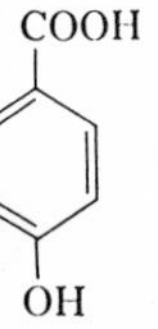

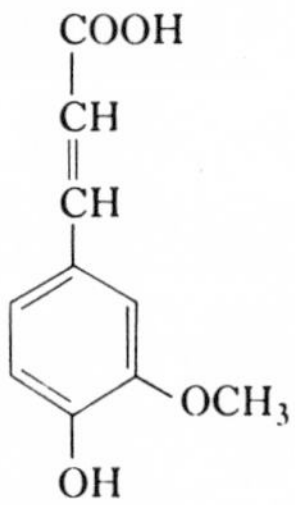

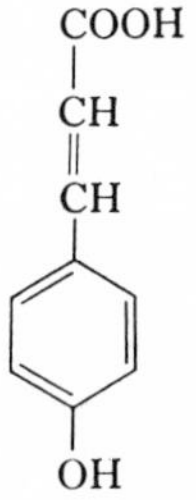

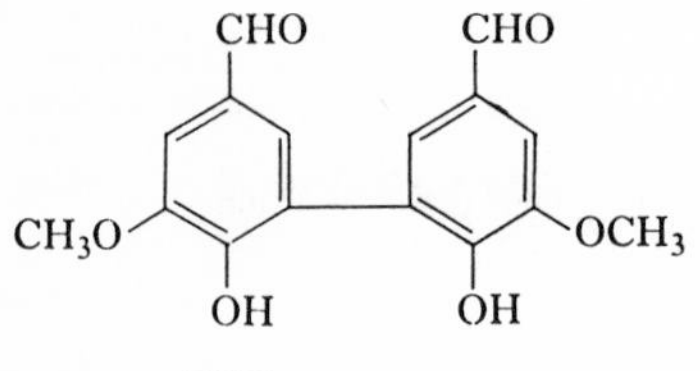

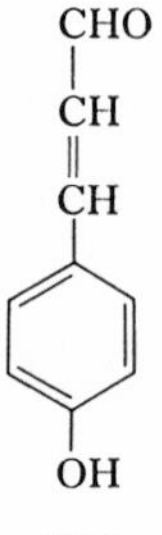

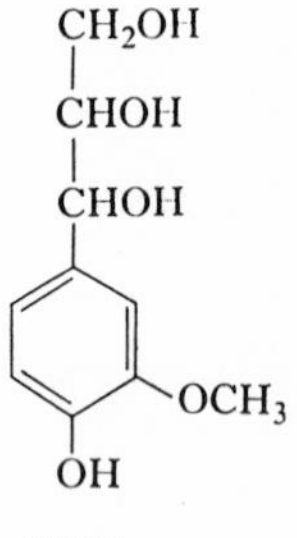

In a subsequent study, paper chromatographic analysis revealed the presence of about fifteen compounds in the ether extracts of pine native and milled-wood lignin-containing media after decomposition by the poly-phenoloxidase-poor fungi, *P. subacida* N199, *F. fomentarius*, *F. annosus*, and *T. pini*. On the other hand, the number of products derived from these lignins after degradation by polyphenoloxidase-rich species, *P. hirsutus*, *P. versicolor*, and *P. subacida* J247 varied from five to seven. The R_f values for the products present in the acidic fraction agreed with the values for vanillic (IV), *p*-hydroxybenzoic (XI), ferulic (XII), and *p*-hydroxycinnamic acids (XIII). The values for the carbonyl compounds corresponded to the enol tautomer of 3-methoxy-4-hydroxyphenylpyruvic acid (IX), vanillin (VII), dehydrodivanillin (XIV), coniferaldehyde (VI), and p-hydroxy-cinnamaldehyde (XV). The values for the phenolic products agreed with those of guaiacylglycerol (XVI) and its β-coniferyl ether (X). The color reactions and positions and intensities of the maximum and minimum points of the ultraviolet absorption spectra of these products all agreed with those of authentic samples (40).

From the pine and spruce lignins degraded by *F. fomentarius*, the formation of the keto and enol tautomers of 3-methoxy-4-hydroxyphenylpyruvic acid (IX), ferulic acid (XII), coniferaldehyde (VI), guaiacylglycerol (XVI), and its β-coniferyl ether (X) was also established by paper chromatographic analysis. However, these products, with the exception of coniferaldehyde, were not found in extracts of the lignin-containing media degraded by *P. versicolor*. These observations indicated that ferulic acid (XII), 3-methoxy-4-hydroxyphenylpyruvic acid (IX), guaiacylglycerol (XVI), and its coniferyl ether (X) were more easily metabolized by polyphenoloxidase-rich fungi than were the other substances. The amounts of degradation products accumulated in a shaken culture medium were usually small (Table I). However, if mycelial pellets were used for inoculation, and the resulting culture was incubated without shaking, the amounts of products were increased three- or four-fold when compared with the amounts accumulated with shaking for the same period of time (40).

The quantities of degradation products of spruce native lignin which accumulated after 28 days of decay by *F. fomentarius* and *C. vellutipes* were also small (Table II). Paper chromatographic analysis established the presence of vanillic acid (IV), vanillin (VII), dehydrodivanillin (XIV), coniferaldehyde (VI), p-hydroxycinnamaldehyde (XV), ferulic acid (XII), 3-methoxy-4-hydroxyphenylpyruvic acid (IX), guaiacylglycerol (XVI),

$$CH_2OH$$
$$|$$
$$C\!-\!OH$$
$$\|$$
$$CH$$

OCH₃

OH

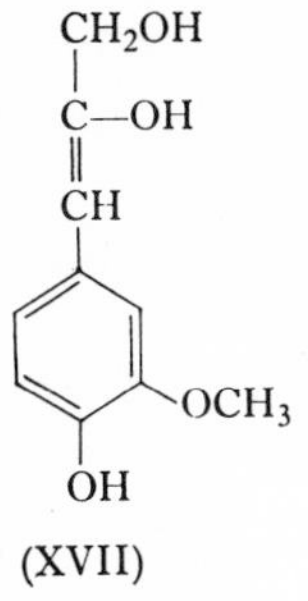

(XVII)

TABLE I

YIELDS OF DEGRADATION PRODUCTS OF PINE LIGNINS

Lignin	Fungus	Vanillic acid (mg.)	p-Hydroxybenzoic acid (mg.)	3-Methoxy-4-hydroxyphenylpyruvic acid (mg.)	p-Hydroxycinnamic acid (mg.)	Guaiacylglycerol (mg.)	Vanillin (mg.)	Coniferaldehyde (mg.)	Guaiacylglycerol-β-coniferyl ether (mg.)
Stationary cultures									
N.L.[a]	F.f.[c]	0.4–0.6	0.4–0.6	Small	0.2–0.4	0.07–0.09	0.3–0.4	0.02–0.03	0.1–0.15
	P.v.[d]	0.2–0.3	0.3–0.5	0.0	0.2–0.3	0.0	0.08–0.11	Small	0.0
M.W.L.[b]	F.f.	0.2–0.3	0.4–0.5	Small	0.3–0.4	0.03–0.05	0.3–0.4	0.02–0.03	0.07–0.1
	P.v.	0.2–0.3	0.2–0.4	0.0	0.2–0.4	0.0	0.05–0.08	Small	0.0
Control		0.0	0.2–0.4	0.0	0.1–0.2	0.0	0.0	0.0	0.0
Shaken cultures									
N.L.	F.f.	0.1–0.2	0.2–0.4	Small	0.1–0.3	Small	0.08–0.11	Trace	Small
	P.v.	0.1–0.2	0.15–0.3	0.0	0.1–0.2	0.0	0.03–0.05	Trace	0.0
Control		0.0	0.13–0.2	0.0	0.07–0.15	0.0	0.0	0.0	0.0

[a] N.L., native lignin.
[b] M.W.L., milled-wood lignin.
[c] F.f., Fomes fomentarius.
[d] P.v., Polyporus versicolor.

TABLE II

YIELDS OF THE DEGRADATION PRODUCTS OF SPRUCE NATIVE LIGNIN

Organism	Vanillic acid[a] (mg.)	3-Methoxy-4-hydroxyphenylpyruvic acid[a] (mg.)	Guaiacylglycerol[a] (mg.)	Vanillin[a] (mg.)	Coniferaldehyde[a] (mg.)	Guaiacylglycerol-β-coniferyl ether[a] (mg.)
Control[b]	0.0	0.0	0.0	0.0	0.0	0.0
Fomes fomentarius	0.7–1.1	Small amount	0.1–0.2	0.6–0.8	0.03–0.05	0.1–0.3
Collybia velutipes	0.5–0.9	Small amount	0.05–0.2	0.4–0.7	0.02–0.06	0.08–0.11

[a] Milligrams of degradation products per 2 gm. lignin as source of carbon.
[b] Lignin-free media after inoculation with *Fomes fomentarius* or *Collybia velutipes*.

guaiacylglycerol-β-coniferyl ether (X), and other unidentified aromatic compounds. The R_f values for one such compound agreed with the reported values for β-hydroxyconiferyl alcohol (XVII) (42).

D. Biochemical Conversions of Lignin Degradation Products

Various isolated lignins have been used in attempted studies of the intermediary phases of lignin decay by the activities of wood-destroying fungi. However, it must not be overlooked that the results obtained are often inconclusive, partly because of the likelihood of chemical alteration in the structure of the lignin occurring during its isolation, and partly because of the inadvertent retention of nonlignin wood components during the isolation and purification of the lignin preparation; this nonlignin material can conceivably be utilized for the metabolic activities of the organism employed for the decay. Accordingly, to gain some understanding of the mode of degradation of lignin and of its metabolic products, many investigations have been performed on the biochemical conversions of certain simpler aromatic compounds, whose structures are believed to be "related" to the structure of lignin, by the enzyme systems of wood-destroying fungi. Ideally, such investigations should involve the conversions of compounds which are known to be enzymically produced, either directly or indirectly, from lignin, and which are obtainable in a pure form. In some instances, however, compounds described merely as "chemically related" to lignin have been used. Obviously, the validity of such investigations is determined directly by the *de facto* chemical identity or nonidentity of the compounds employed with actual metabolic products of lignin. Despite their appeal because of greater availability, *chemical* degradation products of lignin, often obtained by drastic reactions, are obviously not necessarily obligatory intermediates of the natural fungal decay of lignin and observations made on them are not always meaningful.

Studies of this type may be undertaken either by cultivating the lignin-decomposing fungus in a medium containing the "lignin-related" compound as its sole organic source and examining the resultant culture filtrate for its metabolic products, or, alternatively, by incubating the compound with enzyme solutions or preparations previously obtained from either the mycelium of the organism or from the aqueous medium in which it had been cultivated and, again, examining the incubated solution for metabolic products. Studies of all these kinds have been reported.

1. BY LIGNIN-DECOMPOSING FUNGI

In 1952, α-conidendrin (XVIII) was employed (45) as a "model substance" for lignin in microbiological degradation experiments. A *Flavobacterium* was isolated from soil which caused an almost complete degradation of the α-conidendrin in 3 weeks, and simultaneously produced a dark

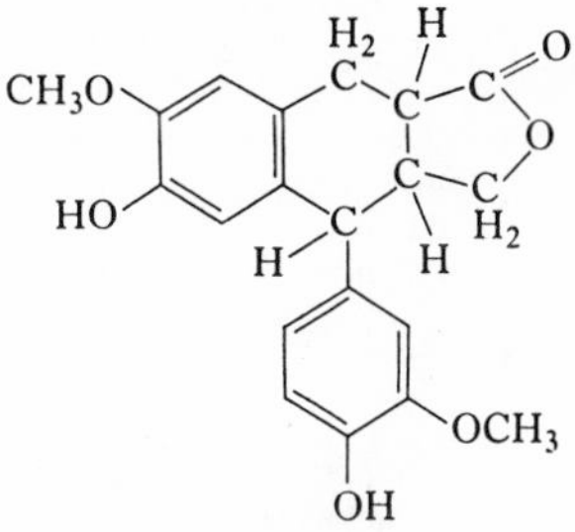

(XVIII)

brown-colored substance. Other methoxylated aromatic compounds were also studied, and it was found that trimethoxybenzoic acid, ferulic acid (XII), and anisyl acetate (XIX) were converted to protocatechuic acid (XX). The isolation of α-conidendrin-decomposing organisms from other sources has

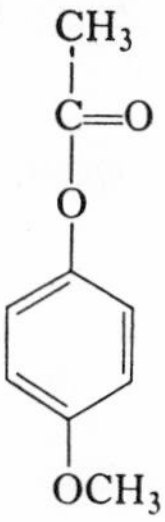

(XIX)

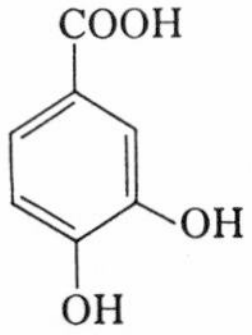

(XX)

also been reported (72, 73). However, one α-conidendrin-decomposing *Flavobacterium* was tested for growth on native lignin and on some thirty "lignin-related" substances, but only vanillic acid (IV) supported growth (65).

Henderson and her co-workers (23) reported that *P. versicolor* produced an extracellular aromatic alcohol oxidase that was capable of oxidizing aromatic alcohols of the *primary* type, such as coniferyl (XXI), vanillyl

(XXII), and veratryl alcohols (XXIII), to the corresponding aldehydes, while aromatic alcohols of the *secondary* type, such as guaiacylglycol (XXIV)

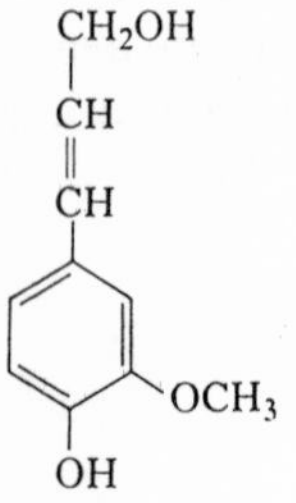

CH$_2$OH
|
CH
‖
CH

OCH$_3$
OH
(XXI)

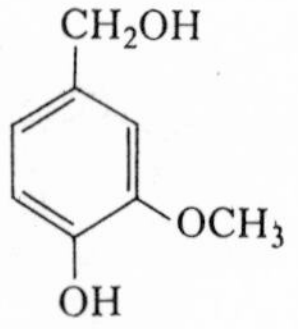

CH$_2$OH

OCH$_3$
OH
(XXII)

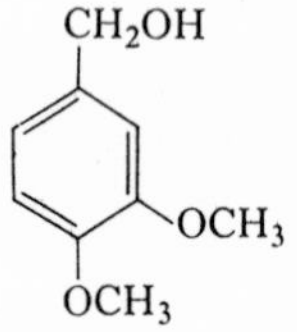

CH$_2$OH

OCH$_3$
OCH$_3$
(XXIII)

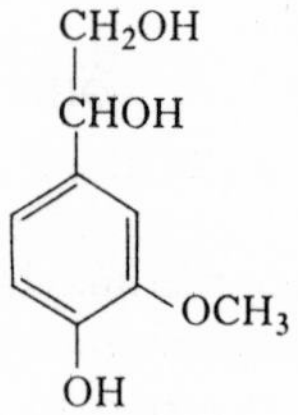

CH$_2$OH
|
CHOH

OCH$_3$
OH
(XXIV)

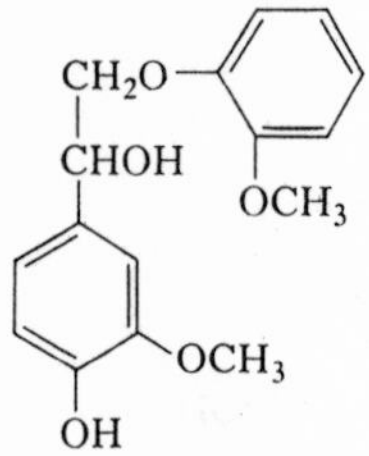

CH$_2$O—
|
CHOH
OCH$_3$

OCH$_3$
OH
(XXV)

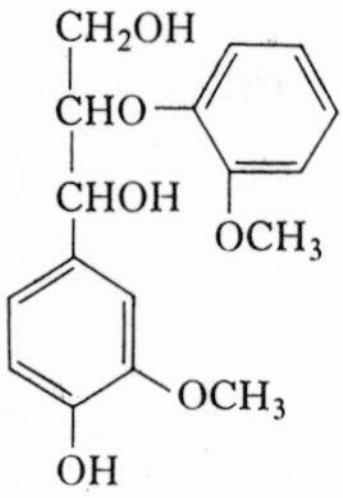

CH$_2$OH
|
CHO—
|
CHOH
OCH$_3$

OCH$_3$
OH
(XXVI)

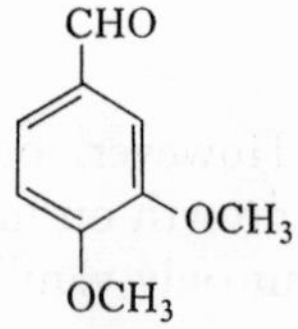

CHO

OCH$_3$
OCH$_3$
(XXVII)

and guaiacylglycerol (XVI), were not oxidizable. They also found (66) that mats of this organism metabolized the β-guaiacyl ethers of guaiacylglycol

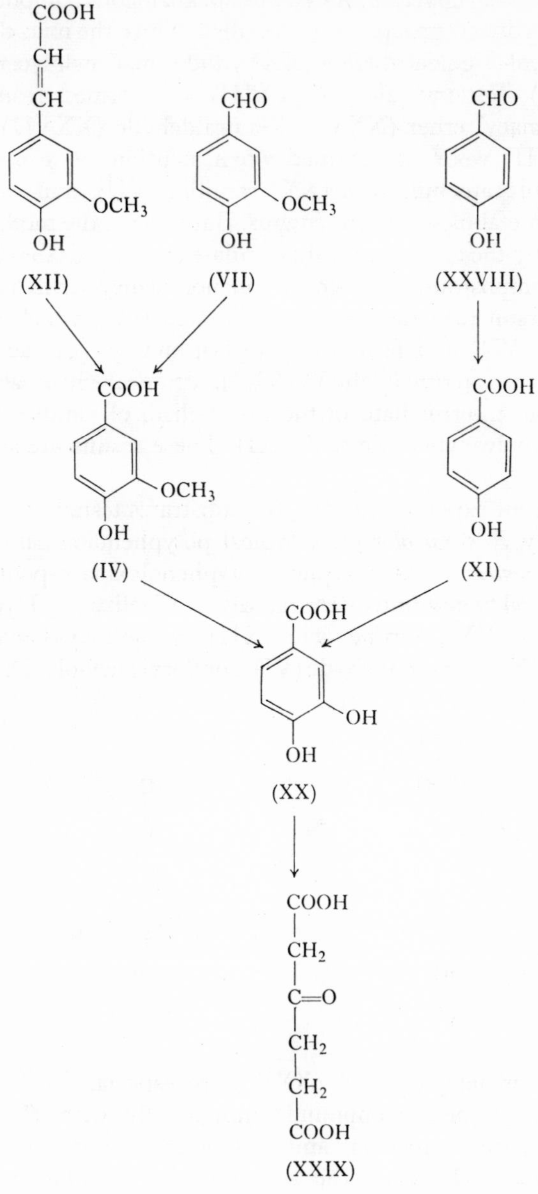

Fig. 4. Decomposition products of phenolic compounds produced by certain soil fungi.

(XXIV) and guaiacylglycerol (XVI), but not analogous compounds in which the phenolic hydroxyl groups were etherified. Thus, the mats did metabolize α-guaiacylglycol-β-guaiacyl ether (XXV) and α-guaiacylglycerol-β-guaiacyl ether (XXVI). Veratryl alcohol (XXIII) was formed from α-guaiacyl-glycerol-β-guaiacyl ether (XXVI). Veratraldehyde (XXVII) and veratryl alcohol (XXIII) were also formed when solutions of p-hydroxybenzoic acid (XI), protocatechuic acid (XX), vanillin (VII), and vanillyl alcohol (XXII) were metabolized by the fungus. But none of the model compounds was oxidized by the aromatic alcohol oxidase of P. versicolor (66).

In the decomposition products of phenolic compounds formed by soil fungi, Henderson (32) identified vanillic acid (IV), which was obtained from vanillin (VII) and ferulic acid (XII), and syringic acid (V), which was formed from syringaldehyde (VIII). Protocatechuic acid (XX) was found to be an intermediate in the metabolism of vanillin (VII), ferulic (XII), and p-hydroxybenzoic acids (XI). These results are summarized in Fig. 4.

In more recent experiments (41), the substrates tested were subjected to metabolism by P. versicolor (as a typical polyphenoloxidase-rich species) and by F. fomentarius (as a typical polyphenoloxidase-poor species). In general, guaiacyl compounds were rapidly metabolized. p-Hydroxyphenyl-pyruvic acid (XXX), ferulic acid (XII), 3-methoxy-4-hydroxyphenyl-pyruvic acid (IX), coniferaldehyde (VI), coniferyl alcohol (XXI), isoeugenol

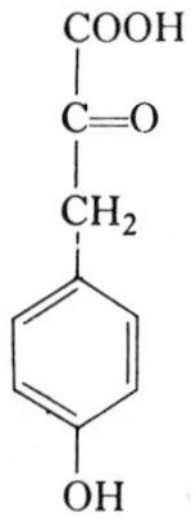

COOH
|
C=O
|
CH₂

OH

(XXX)

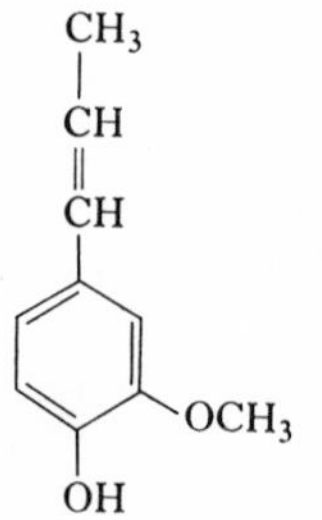

CH₃
|
CH
‖
CH

OCH₃
OH

(XXXI)

(XXXI), and guaiacylglycerol (XVI) were especially well metabolized. After incubating these compounds individually with P. versicolor for 4–5 days, only small residual amounts of them were detectable in the resulting filtrates. However, the amounts of metabolic products which formed and accumulated in the media were also small (Table III).

TABLE III

YIELDS OF PRINCIPAL DEGRADATION PRODUCTS FORMED FROM AROMATIC
COMPOUNDS BY *Fomes fomentarius* AND *Polyporus versicolor*[a]

Substrates	Period of incubation (days)	Recovered substrate (%)	Degradation products	
			p-Hydroxy-benzaldehyde (%)	p-Hydroxy-benzoic acid (%)
A. *Fomes fomentarius*				
p-Hydroxybenzoic acid	2	78.2	—	—
	5	41.3	—	—
p-Hydroxybenzaldehyde	2	90.0	—	7.2
	5	78.1	—	16.5
p-Hydroxycinnamic acid	1	63.8	15.9	16.4
	2	26.5	5.0	23.6
	5	16.1	Trace	28.4
p-Hydroxyphenylpyruvic acid	1	58.2	25.1	10.7
	2	20.6	32.7	18.2
	5	9.5	24.4	8.5

Substrates	Period of incubation (days)	Recovered substrate (%)	Degradation products	
			Vanillin (%)	Vanillic acid (%)
Vanillic acid	2	60.1	—	—
	5	38.6	—	—
Vanillin	1	91.1	—	2.8
	2	61.4	—	21.5
	5	36.6	—	18.9
Vanillyl alcohol	2	80.3	13.7	9.7
	5	52.8	15.6	14.2
Acetovanillone	2	60.7	12.5	7.7
	5	40.6	8.3	11.2
Guaiacylmethyl carbinol	2	83.5	3.0	3.3
	5	51.1	7.5	8.4
Ferulic acid	1	40.2	7.1	1.8
	2	20.0	11.5	8.5
	5	10.5	5.9	23.3
3-Methoxy-4-hydroxy-phenylpyruvic acid	1	28.4	13.7	4.4
	2	19.0	22.6	15.1
	5	6.6	8.3	14.8

[a] See footnote, page 104.

TABLE III—*cont.*

| | | | Degradation products | |
| | Period of incubation (days) | Recovered substrate (%) | *p*-Hydroxy-benzaldehyde (%) | *p*-Hydroxy-benzoic acid (%) |
Substrates				
B. *Polyporus versicolor*				
p-Hydroxybenzoic acid	2	85.0	—	—
	5	40.0	—	—
p-Hydroxybenzaldehyde	2	80.4	—	11.6
	5	36.3	—	22.3
p-Hydroxycinnamic acid	1	53.3	11.3	10.6
	2	20.0	9.7	21.2
	5	13.3	5.0	19.1
p-Hydroxyphenylpyruvic acid	1	39.6	22.5	4.9
	2	13.3	15.8	9.7
	5	5.8	7.0	4.0

| | | | Degradation products | |
| | Period of incubation (days) | Recovered substrate (%) | Vanillin (%) | Vanillic acid (%) |
Substrates				
Vanillic acid	2	49.6	—	—
	5	20.3	—	—
Vanillin	1	74.8	—	5.0
	2	32.6	—	14.4
Vanillyl alcohol	2	38.2	10.2	8.5
	5	18.0	11.7	12.6
Acetovanillone	1	86.0	2.8	1.8
	5	30.2	5.6	4.8
Guaiacylmethyl carbinol	5	35.5	6.2	5.7
Ferulic acid	1	22.5	9.6	7.9
	2	13.1	17.8	13.7
	5	5.1	4.4	—
3-Methoxy-4-hydroxy-phenylpyruvic acid	1	20.7	17.9	5.3
	2	11.0	18.6	11.6
	5	4.0	5.5	10.6

[a] Initial concentration: 0.1 gm. in 200 ml. of 1% glucose-containing medium.

A certain amount of all of the aromatic compounds tested was converted to a colored polymer, apparently by the activity of an extracellular polyphenoloxidase. *p*-Hydroxycinnamic (XIII), *p*-hydroxyphenylpyruvic (XXX), ferulic (XII), and 3-methoxy-4-hydroxyphenylpyruvic (IX) acids,

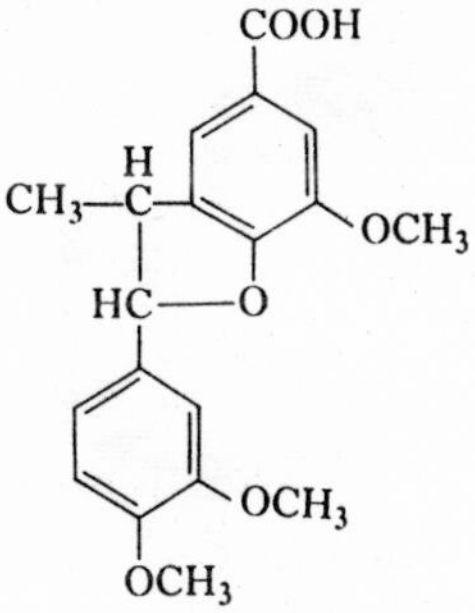

(XXXII)

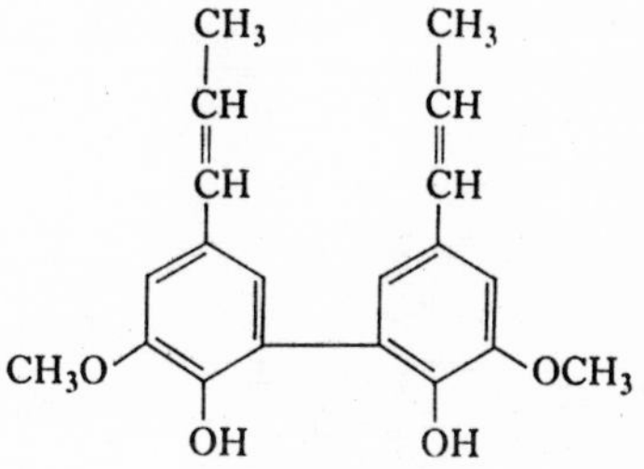

(XXXIII)

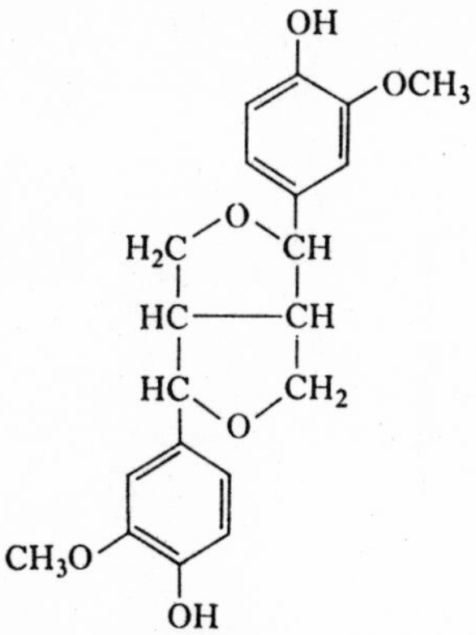

(XXXIV)

Erdtman's acid (XXXII), coniferaldehyde (VI), coniferyl alcohol (XXI), isoeugenol (XXXI), dehydrodiisoeugenol (XXXIII), symplocosigenol (XXXIVa),[1] and pinoresinol (XXXIVb) were more easily polymerized by the polyphenoloxidase-rich *P. versicolor* than by the enzyme-poor *F. fomentarius*.

[1] Compounds (XXXIVa) and (XXXIVb) are optical isomers of structure (XXXIV).

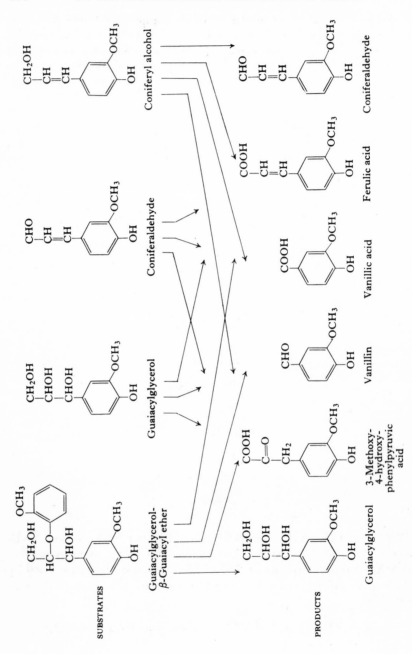

Fig. 5.   Conversions of substrates by pellet suspensions of *P. versicolor* and *F. fomentarius.*

In one series of experiments, guaiacylglycerol-β-guaiacyl ether, guaiacyl-glycerol, coniferaldehyde, coniferyl alcohol, and certain other compounds were employed as substrates, and were incubated in the presence of suspensions of pellets of *P. versicolor* and *F. fomentarius*. The natures of the conversions and the identities and structures of the products formed from the first four of these substrates are shown in Fig. 5.

After incubating the substrates with pellets of the organisms for 4–5 days, only small amounts of the substrates remained in the filtrates. But the amounts of metabolic products formed from these compounds and accumulated in the media were also small. The quantitative data for all the substrates tested are presented in Table IV.

After the degradation of several guaiacyl compounds by *F. fomentarius*, a compound was detected which was identified as vanillyl alcohol (XXII). Furthermore, small amounts of *p*-hydroxybenzoic acid (XI), *p*-hydroxy-benzaldehyde (XXVIII), *p*-hydroxycinnamic acid (XIII), and ferulic acid (XII) were enzymically converted to the corresponding aldehyde or alcohol (Table V).

The demethoxylation of mono- and dimethoxybenzoic acids by both *F. fomentarius* and *P. versicolor* was indicated by the formation of small amounts of *p*-hydroxybenzoic (XI) and vanillic (IV) acids. The methyl derivatives of other guaiacyl compounds studied were partially demethoxyl-ated by *F. fomentarius*, and these products were further oxidized to vanillic acid (IV) and vanillin (VII), but the yields were small (Table VI) (41).

2. By Mycelium-Free Culture Filtrates

Mycelium-free culture filtrates of *F. fomentarius* and *C. velutipes* were obtained by the filtration of cultures of the organisms which had previously been grown, with shaking, on a glucose-containing medium (42). Portions of the filtrate were then incubated without shaking for 2 days after the addition to them of either guaiacylglycerol (XVI), β-hydroxyconiferyl alcohol (XVII), or 3-methoxy-4-hydroxyphenylpyruvic acid (IX). After extracting the resulting solutions with ether, their constituents were separated into carbonyl, carboxyl, phenolic, and neutral fractions. The principal products present in each extract were then determined by paper chromatography. The identities of the conversion products of these substrates are shown in Fig. 6. The quantitative data for these experiments are given in Table VII.

These observations indicated that guaiacylglycerol (XVI), presumably derived from the guaiacylglycerol-β-aryl ether units of the lignin, was

TABLE IV

YIELDS OF PRINCIPAL DEGRADATION PRODUCTS FORMED FROM GUAIACYL COMPOUNDS BY WHITE-ROT FUNGI

Substrates	Period of incubation (days)	Recovered substrate (%)	Degradation products (%)					
			Vanillic acid	Vanillin	Ferulic acid	Coniferaldehyde	3-Methoxy-4-hydroxyphenylpyruvic acid	Guaiacylglycerol
Fomes fomentarius								
Coniferaldehyde	2	11.6	2.02	3.26	Small amount	—	—	—
Coniferyl alcohol	2	9.2	1.64	2.77	Trace	Small amount	—	—
	4	1.2	2.48	1.99	Trace	Trace	—	—
Isoeugenol	2	40.1	1.38	2.05	Trace	Small amount	—	—
	4	5.2	3.07	2.33	Trace	Small amount	—	—
Guaiacylglycerol	2	25.1	1.47	2.93	—	—	0.21	—
	4	17.6	3.40	1.53	—	—	Trace	—
Dehydrodiisoeugenol	2	85.1	0.78	1.14	—	—	—	—
	5	62.8	1.00	0.79	—	—	—	—
Erdtman's acid	5	63.0	0.40	1.22	—	—	—	—
Symplocosigenol	5	78.0	1.10	0.68	—	—	—	—
Pinoresinol	5	59.8	1.52	0.43	—	—	—	—
Guaiacylglycerol-β-guaiacyl ether	2	75.9	1.56	3.01	—	—	Small amount	0.15
	4	33.5	1.85	2.30	—	—	Small amount	0.87
Polyporus versicolor								
Coniferaldehyde	2	5.8	1.66	2.18	Trace	—	—	—
Coniferyl alcohol	2	5.5	1.70	2.09	Trace	Small amount	—	—
Isoeugenol	2	28.0	1.44	0.92	—	Small amount	—	—
Guaiacylglycerol	2	8.0	0.61	1.72	—	—	Trace	—
Dehydrodiisoeugenol	5	36.0	0.54	0.65	—	—	—	—
Erdtman's acid	5	32.0	0.82	0.78	—	—	—	—
Pinoresinol	5	43.6	0.64	0.57	—	—	—	—
Symplocosigenol	5	56.0	0.55	0.53	—	—	—	—
Guaiacylglycerol-β-guaiacyl ether	4	17.9	1.87	1.11	—	—	Trace	—

TABLE V

REDUCTION PRODUCTS FORMED FROM AROMATIC
COMPOUNDS BY *Fomes fomentarius*

Substrates	Period of incubation (days)	Reduction products
Vanillin, vanillic acid, acetovanillone, and isoeugenol	4	Vanillyl alcohol
Ferulic acid	2	Vanillyl alcohol and coniferaldehyde
p-Hydroxybenzoic acid	4	*p*-Hydroxybenzaldehyde and *p*-hydroxybenzyl alcohol
p-Hydroxybenzaldehyde	4	*p*-Hydroxybenzyl alcohol
p-Hydroxycinnamic acid	2	*p*-Hydroxycinnamaldehyde

TABLE VI

PRINCIPAL DEGRADATION PRODUCTS FORMED FROM
METHYLGUAIACYL COMPOUNDS BY *Fomes fomentarius*

Substrates	Period of incubation (days)	Degradation products
p-Methoxybenzoic acid	4	*p*-Hydroxybenzoic acid
Veratric acid	4	Vanillic acid
Veratraldehyde, veratryl alcohol, acetovanillone, and 3,4-dimethoxycinnamic acid	4	Vanillin and vanillic acid
3,4-Dimethoxycinnamyl alcohol and aldehyde	3	Vanillin, vanillic acid, and coniferaldehyde
Veratrylglycerol	4	Vanillin, vanillic acid, and guaiacylglycerol (trace)
Veratrylglycerol-β-guaiacyl ether	4	Vanillin, vanillic acid, and guaiacylglycerol-β-guaiacyl ether (trace)

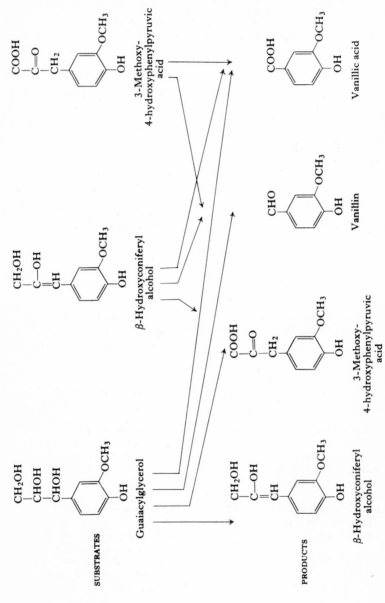

Fig. 6. Conversions of substrates by mycelium-free culture filtrates of *F. fomentarius* and *C. velutipes*.

converted to vanillic acid (IV) via β-hydroxyconiferyl alcohol (XVII), 3-methoxy-4-hydroxyphenylpyruvic acid (IX), and vanillin (VII) by extracellular enzymes secreted into the culture medium by the growing organism (42).

TABLE VII

DEGRADATION OF AROMATIC COMPOUNDS IN THE MYCELIA-FREE CULTURE
FILTRATES OF *Fomes fomentarius* AND *Collybia velutipes*

Substrates	Fungus	Vanillic acid[a] (mg.)	Vanillin[a] (mg.)	3-Methoxy-4-hydroxyphenyl-pyruvic acid	β-Hydroxyconiferyl alcohol-like compound
Control[b]		0.0	0.0	0.0	0.0
3-Methoxy-4-hydroxy-	*Fomes fomentarius*	7.2	4.6	—	—
phenyl pyruvic acid	*Collybia velutipes*	5.0	3.5	—	—
β-Hydroxyconiferyl	*Fomes fomentarius*	5.6	4.9	Small amount	—
alcohol	*Collybia velutipes*	3.6	2.0	Small amount	—
Guaiacylglycerol	*Fomes fomentarius*	0.7	0.5	Small amount	Small amount
	Collybia velutipes	0.4	0.3	Small amount	Small amount

[a] Milligrams of degradation products per 0.5 gm. of substrate in 400 ml. of filtrate.
[b] Filtrate without substrate.

3. BY ISOLATED ENZYMES

Mason and Cronyn (61) have presented evidence which indicates that the polymerization of coniferyl alcohol is not catalyzed by an extracellular polyphenoloxidase. They suggested that the active enzyme may be a laccase. Malmström *et al.* (59) found that *P. versicolor* could be induced to form large amounts of an exoenzyme possessing laccase-like activity. Schneider (67) claimed to have separated tyrosinase (phenolase) and four laccase-like enzymes from a crude preparation isolated from the mushroom *P. campestris*.

The degradation of guaiacylglycerol (XVI), β-hydroxyconiferyl alcohol (XVII), and 3-methoxy-4-hydroxyphenylpyruvic acid (IX) was also achieved by enzyme preparations obtained from the mycelial pellets and mycelium-free culture filtrates of *F. fomentarius* and *C. velutipes*. The methods of preparation and purification of these enzyme solutions have been described (42).

After incubation of the substrates with the purified enzyme solutions, the principal products formed were vanillin (VII), vanillic acid (IV), and oxalic

TABLE VIII

DEGRADATION OF AROMATIC COMPOUNDS BY ENZYME PREPARATIONS

A. "Fungal Oxidases" Isolated from Mycelial Pellets and Culture Filtrates

	Fomes fomentarius				Collybia velutipes					
	Mycelial pellets		Culture filtrate		Mycelial pellets				Culture filtrate	
					I*		II*		I*	
Substrates	a (mg.)	b (mg.)	a (mg.)	b (mg.)	a (mg.)	b (mg.)	a (mg.)	b (mg.)	a (mg.)	b (mg.)
3-Methoxy-4-hydroxyphenyl-pyruvic acid	3.52 (5.11)[c]	3.29 (4.46)	4.55 (7.05)	3.01 (3.78)	0.36 (1.11)	0.44 (1.62)	0.19 (0.31)	0.12 (0.25)	1.03 (2.63)	0.89 (1.99)
β-Hydroxyconiferyl alcohol	0.87	0.88	1.50	1.23	0.31 (0.65)	0.27 (0.59)	—	—	0.31	0.40
Guaiacylglycerol	0.15 (0.33)	0.28 (0.35)	0.34 (0.51)	0.44 (0.49)	0.08 (0.14)	0.13 (0.19)	Small (0.05)	Small (0.01)	Small (0.33)	0.12 (0.28)

B. Phenol Oxidases and Peroxidase

| | Horse radish peroxidase | | | Mushroom tyrosinase | | | Mushroom laccase | | |
Substrates	d (mg.)	a (mg.)	b (mg.)	d (mg.)	a (mg.)	b (mg.)	d (mg.)	a (mg.)	b (mg.)
3-Methoxy-4-hydroxyphenylpyruvic acid	13.6	Trace (5.01)	Trace (4.27)	40.00	Trace	Trace	19.3	2.27	3.51
β-Hydroxyconiferyl alcohol	24.4	Trace (2.99)	Trace (3.00)	42.2	Trace	Trace	28.5	2.30	2.88
Guaiacylglycerol	30.00	Trace	Trace	46.8	Trace	Trace	35.1	1.57	1.36

* I, Peroxidase-containing enzyme fraction. II, Enzyme fraction containing no peroxidase.
[a] Vanillic acid.
[b] Vanillin.
[c] The values shown in parentheses were obtained upon addition of 0.03 M (final concentration) hydrogen peroxide to each enzyme solution.
[d] Milligrams of recovered substrate.

acid. The formation of vanillin was accelerated by the addition of a small amount of hydrogen peroxide to the solutions, and the rate of oxidation of the substrates was accelerated by increasing the concentration of enzyme. Heating at 100°C. for 15 minutes completely inhibited the reaction (42).

Paper chromatographic analysis revealed that 3-methoxy-4-hydroxyphenylpyruvic acid (IX) was formed from β-hydroxyconiferyl alcohol (XVII) and guaiacylglycerol (XVI), and a β-hydroxyconiferyl alcohol-like compound was also obtained from guaiacylglycerol. The relative amounts of products accumulated were greater in enzyme-poor solutions than they were in enzyme-rich solutions.

The activities on the above substrates of the enzyme solutions obtained from *F. fomentarius* and *C. velutipes* were compared with the analogous activities of a "fungal laccase" obtained from the mushroom *P. campestris* and of preparations of peroxidase and phenolase obtained from purified commercial horse radish peroxidase and from "mushroom tyrosinase," respectively (Table VIII) (42).

These experiments demonstrated that the conversions of 3-methoxy-4-hydroxyphenylpyruvic acid (IX), β-hydroxyconiferyl alcohol (XVII), and guaiacylglycerol (XVI) to vanillin (VII), vanillic acid (IV), and oxalic acid were catalyzed by the fungal enzymes, by laccase, and by peroxidase in the presence of hydrogen peroxide, but not by phenolase. The formation of 3-methoxy-4-hydroxyphenylpyruvic acid (IX) from β-hydroxyconiferyl alcohol (XVII) or guaiacylglycerol (XVI) and of a comparatively large amount of a β-hydroxyconiferyl alcohol-like compound from guaiacylglycerol (XVI) was also established. A small amount of all the substrates tested was converted to a polymer by the laccase and by peroxidase in the presence of hydrogen peroxide (42).

From the identity of the degradation products of guaiacylglycerol and of β-hydroxyconiferyl alcohol, it was concluded that guaiacylglycerol is converted to vanillin, vanillic and oxalic acids via β-hydroxyconiferyl alcohol and 3-methoxy-4-hydroxyphenylpyruvic acid, by the activities of the fungal enzymes. It, therefore, appears that guaiacylglycerol, β-hydroxyconiferyl alcohol, and 3-methoxy-4-hydroxyphenylpyruvic acid are intermediates in the enzymic degradation of the guaiacylglycerol-β-aryl ether units of softwood lignin and that 3-methoxy-4-hydroxyphenylpyruvic acid is converted to vanillin, vanillic acid, and oxalic acid by extracellular oxidase enzymes of the wood-destroying fungi studied.

IV. CONCLUSION

As indicated in the first chapter, it is generally agreed that softwood lignin is derived from phenylpropanoid compounds, such as coniferyl alcohol, and that certain of its building stones contain a benzyl alcohol structure with a free phenolic hydroxyl group at the position *para* to the side chain (12, pp. 227, 616). A color reaction for the detection of the coniferyl alcohol structure in lignin was developed by Lindgren and Mikawa (56). The presence of one coniferaldehyde group per 20–40 phenylpropane-building stones has been demonstrated by several investigators (2). These observations suggest the possibility of the formation of products of the type of coniferaldehyde, ferulic acid and guaiacylglycerol from lignin as the result of fungal or enzymic attack, and this has in fact been realized.

The presence of small amounts of carbonyl groups and of double bonds conjugated with phenolic rings in softwood lignin has also been demonstrated by many investigators (2, 3, 27). These observations are consistent

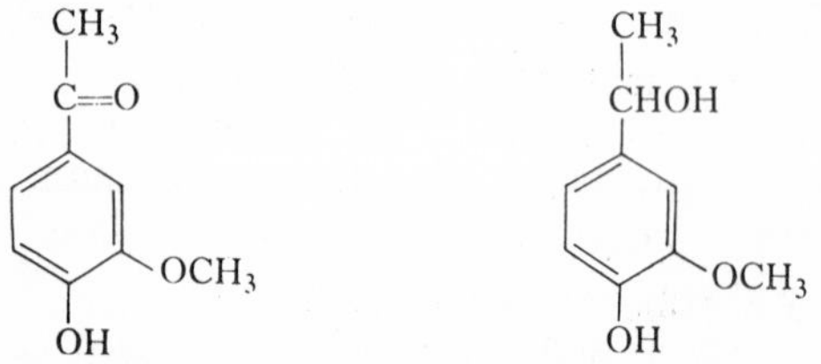

with the presence of coniferyl alcohol, coniferaldehyde, and ferulic acid structures, and small amounts of these have also been detected in softwood lignin (12, p. 265; 69). Experiments with low molecular weight lignin "model compounds" indicate the possibility of the oxidation of such moieties to vanillin and vanillic acid. The alkaline nitrobenzene oxidation of softwood lignins and of "model compounds" has indicated that the vanillin-yielding moieties present in such lignins include structures of the type of guaiacyl ketones, guaiacyl alcohols, and ethers of the latter (46, 50). Thus, vanillyl alcohol (XXII), acetovanillone (XXXV), guaiacylmethyl carbinol (XXXVI), coniferyl alcohol (XXI), coniferaldehyde (VI), iso-eugenol (XXXI), ferulic acid (XII), and guaiacylglycerol (XVI) are all rapidly oxidized to vanillin and vanillic acid under these conditions. Obviously, double bonds conjugated with a phenolic ring are easily oxidized by this treatment. But it has been repeatedly observed that the yields of vanillin obtainable from lignins after their decay by white-rot fungi are lower than the yields obtainable from the original undecomposed lignins (40). These observations clearly indicate that the vanillin-yielding moieties of softwood lignin are preferentially attacked by white-rot fungi.

From the results of the acidolysis of lignin and of guaiacylglycerol-β-guaiacyl ether (XXVI), Adler and co-workers (3) concluded that from one-third to one-fourth of the building stones of softwood lignin may be guaiacyl-glycerol moieties carrying β-aryl ether units. The presence of guaiacyl-glycerol (XVI) and guaiacylglycerol-β-coniferyl ether (X) in the products of the fungal degradation of pine and spruce native lignins has now been established (40). These observations are, therefore, completely consistent with the presence in these lignins of a guaiacylglycerol-β-coniferyl ether unit as an integral part of their structures.

The possible occurrence of phenylcoumaran and of pinoresinol structures in softwood lignin has also been suggested (1, 3). But in experiments with "model compounds", Erdtman's acid (XXXII), dehydrodiisoeugenol (XXXIII), pinoresinol (XXXIVb), and symplocosigenol (XXXIVa) were relatively resistant to degradation by the white-rot fungi employed (41). These observations would seem to imply that moieties of the phenyl-coumaran and pinoresinol types, if they are indeed present in lignin, are more resistant to degradation by the fungi studied than are other parts of the structure.

The initial attack by white-rot fungi on softwood lignins obviously is made by extracellular enzymes, and results in a loss of methoxyl groups and in the cleavage of certain other ether linkages of the lignin, thereby causing a partial degradation of the lignin to more soluble substances and a removal of some of its "vanillin-yielding" moieties. Low molecular weight products derived in this way have been identified as coniferaldehyde, p-hydroxy-cinnamaldehyde, ferulic acid, 3-methoxy-4-hydroxyphenylpyruvic acid, p-hydroxycinnamic acid, guaiacylglycerol and the β-coniferyl ether of the latter. The formation of vanillin, dehydrodivanillin, and vanillic acid was also established. As already mentioned, these results are consistent with the presence of guaiacylglycerol-β-coniferyl ether units in softwood lignins.

It further appears that these guaiacylglycerol moieties carrying β-aryl ether units are converted to 3-methoxy-4-hydroxyphenylpyruvic acid, vanillin, and vanillic acid. (Coniferaldehyde and ferulic acid, also derived from softwood lignins by fungal degradation, are likewise converted to vanillin and vanillic acid.) Furthermore, it appears that guaiacylglycerol, β-hydroxyconiferyl alcohol, and 3-methoxy-4-hydroxyphenylpyruvic acid

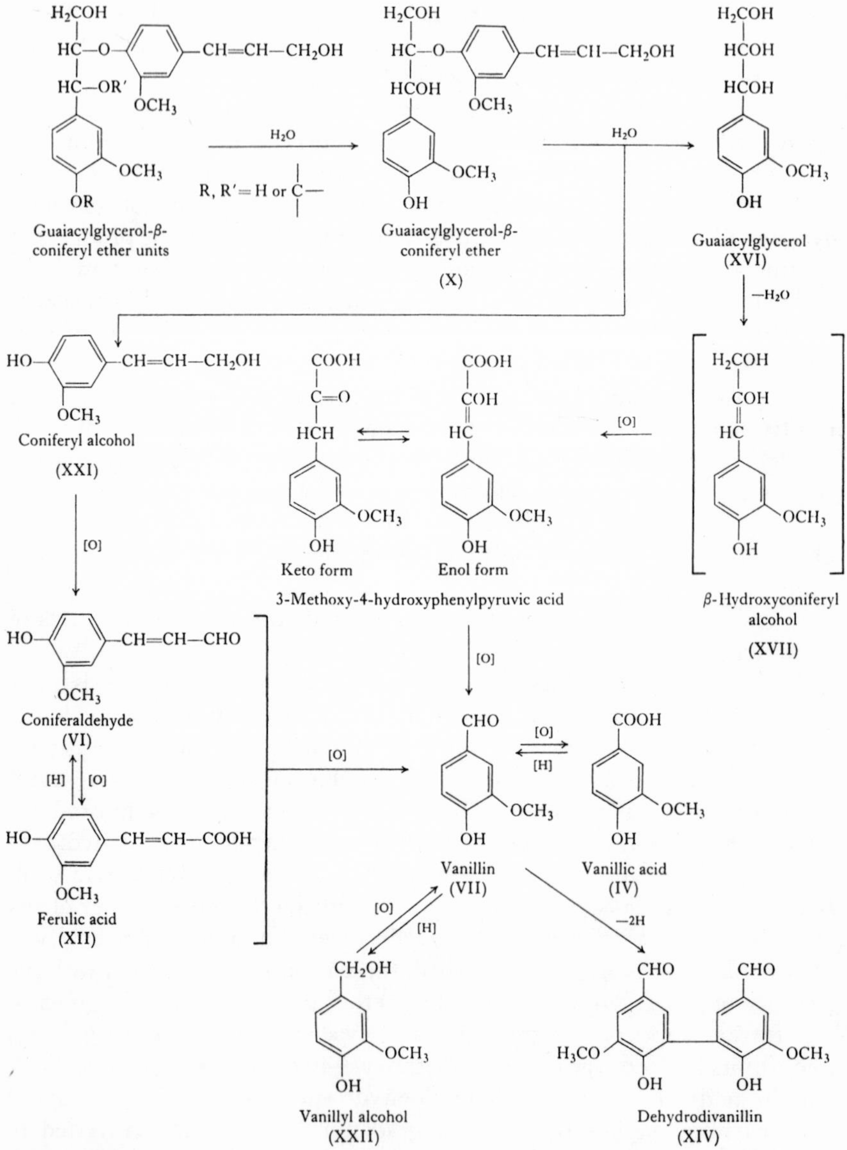

H_2O

R, R′ = H or C—

Guaiacylglycerol-β-
coniferyl ether units

H_2O

Guaiacylglycerol-β-
coniferyl ether

(X)

Guaiacylglycerol
(XVI)

$-H_2O$

Coniferyl alcohol

(XXI)

[O]

COOH
C=O
HCH

Keto form

COOH
COH
HC

Enol form

3-Methoxy-4-hydroxyphenylpyruvic acid

[O]

H_2COH
COH
HC

β-Hydroxyconiferyl
alcohol

(XVII)

Coniferaldehyde
(VI)

[H] [O]

[O]

Ferulic acid
(XII)

[O]

[O]

CHO

Vanillin
(VII)

[O]

[H]

COOH

Vanillic acid
(IV)

[O]

[H]

CH_2OH

Vanillyl alcohol
(XXII)

$-2H$

CHO   CHO

H_3CO

Dehydrodivanillin
(XIV)

Fig. 7. A scheme for the enzymic degradation of guaiacylglycerol-β-coniferyl ether units present in softwood lignin by white-rot fungi.

are intermediates in the enzymic degradation of the guaiacylglycerol-β-aryl ether units in softwood lignins. Finally, it would appear that 3-methoxy-4-hydroxyphenylpyruvic acid is converted to vanillin, vanillic acid, and oxalic acid, by an oxidative cleavage of its propyl side chain. These conversions are summarized in Fig. 7.

Since coniferyl alcohol, coniferaldehyde, and ferulic acid (all of which have ring-conjugated double bonds) are extremely sensitive to oxidase activity, it is likely that the enol form of 3-methoxy-4-hydroxyphenylpyruvic acid (with the same conjugation) would be more easily oxidized by the enzyme than the keto form; furthermore, the enol form, derivable from guaiacylglycerol by dehydration and oxidation, could undergo polymerization in the medium of a polyphenoloxidase-rich organism. Accordingly, it is believed that 3-methoxy-4-hydroxyphenylpyruvic acid is an intermediate in the fungal degradation of the guaiacylglycerol-β-coniferyl ether units of softwood lignin, and that the enol form of this acid is further degraded to vanillin and vanillic acid by a shortening of its side chain. The terminal two-carbon unit may be converted to oxalic acid.

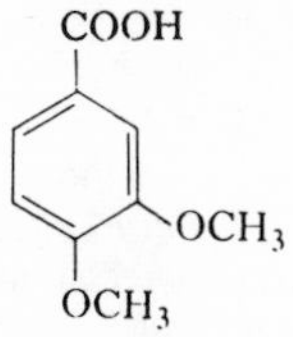

COOH

OCH$_3$

OCH$_3$

(XXXVII)

It has been reported (23, 31, 66) that *P. versicolor* and several "soil fungi" are capable of the demethoxylation and degradation of the methyl derivatives of certain aromatic compounds. Furthermore, the formation of protocatechuic acid (XX) from benzoic, *m*- and *p*-hydroxybenzoic, vanillic, and veratric (XXXVII) acids by these soil fungi has also been reported (23, 66, 70, 73). Protocatechuic acid is well known as an intermediate in the degradation of various aromatic compounds by microorganisms (15, 60). It has now been found that the methyl derivatives of certain aromatic compounds are slowly converted to vanillin and vanillic acid via the corresponding phenolic compounds. But the yields of the products from these methyl derivatives are small when compared with the yields from the corresponding guaiacyl compounds having a free phenolic hydroxyl group (41).

REFERENCES

1. E. Adler, S. Delin, and K. Lundqvist, *Acta Chem. Scand.* **13**, 2149 (1959).
2. E. Adler and J. Marton, *Acta Chem. Scand.* **13**, 75 (1959); **15**, 357, 370 (1961).
3. E. Adler, J. M. Pepper, and E. Eriksoo, *Ind. Eng. Chem.* **49**, 1391 (1957).
4. A. Apenitis, H. Erdtman, and B. Leopold, *Svensk. Kem. Tidskr.* **63**, 195 (1951).
5. J. B. Bartlett and A. G. Norman, *Soil Sci. Soc. Am. Proc.* **3**, 210 (1938).
6. W. Bavendamm, *Zentr. Bakteriol. Parasitenk. Abt. II* **75**, 426 (1928); **76**, 172 (1928).
7. G. Bertrand, *Compt. Rend.* **118**, 1215 (1894); **120**, 266 (1895).
8. G. Bertrand, *Compt. Rend.* **123**, 463 (1896).
9. G. Bertrand and E. Bourquelot, *Compt. Rend.* **121**, 166 (1895); **122**, 1215 (1896).
10. S. R. Bose and S. N. Sarkar, *Proc. Roy. Soc.* (*London*) *Ser. B*, **123**, 193 (1937).
11. F. E. Brauns, "The Chemistry of Lignin." Academic Press, New York, 1952.
12. F. E. Brauns and D. A. Brauns, "The Chemistry of Lignin: Supplement Volume." Academic Press, New York, 1960.
13. K. St. G. Cartwright and W. P. K. Findlay, "Decay of Timber and Its Prevention," 2nd ed., H.M. Stationery Office, London, 1958.
14. F. Czapek, *Ber. Deut. Botan. Ges.* **17**, 166 (1899).
15. S. Dagley, *Nature* **188**, 560 (1960).
16. R. W. Davidson, W. A. Campbell, and D. J. Blaisdell, *J. Agr. Res.* **57**, 683 (1938).
17. C. R. Dawson and W. B. Tarpley, *in* "The Enzymes" (J. B. Sumner and K. Myrbäck, eds.), Vol. II, Part 1, p. 455. Academic Press, New York, 1951.
18. W. C. Day, M. J. Pelczar, and S. Gottlieb, *Arch. Biochem.* **23**, 360 (1949).
19. W. M. Dion, *Can. J. Botany* **30**, 9 (1952).
20. T. Enkvist, E. Solin, and U. Maunula, *Paperi Puu* **36**, 65, 86 (1954).
21. G. Fåhraeus and G. Lindeberg, *Physiol. Plantarum* **6**, 150 (1953).
22. R. Falck and W. Haag, *Ber. Deut. Chem. Ges.* **60**, 225 (1927).
23. V. C. Farmer, M. E. K. Henderson, and J. D. Russel, *Biochem. J.* **74**, 257 (1960).
24. T. Fukuzumi, *Nippon Ringaku Kaishi* **35**, 139 (1953).
25. T. Fukuzumi, *Bull. Agr. Chem. Soc. Japan* **24**, 728 (1960).
26. J. A. Gascoigne and M. M. Gascoigne, "Biological Degradation of Cellulose." Butterworths, London, 1960.
27. J. Gierer and S. Söderberg, *Acta Chem. Scand.* **13**, 137 (1959).
28. S. Gottlieb, W. C. Day, and M. J. Pelczar, *Phytopathology* **40**, 926 (1950).
29. S. Gottlieb and M. J. Pelczar, *Bacteriol. Rev.* **15**, 55 (1951).
30. M. E. K. Henderson, *Nature* **175**, 634 (1955).
31. M. E. K. Henderson, *J. Gen. Microbiol.* **26**, 149 (1961).
32. M. E. K. Henderson, *J. Gen. Microbiol.* **26**, 155 (1961).
33. M. E. K. Henderson, "The Ecology of Soil Fungi," p. 286. Liverpool Univ. Press.
34. T. Higuchi, *Nippon Ringaku Kaishi* **35**, 77 (1953).
35. T. Higuchi, *Nippon Ringaku Kaishi* **36**, 22 (1954).
36. T. Higuchi and H. Kawamura, *Nippon Ringaku Kaishi* **37**, 298 (1955).
37. T. Higuchi, I. Kawamura, and H. Kawamura, *Nippon Ringaku Kaishi* **37**, 547 (1955).
38. T. Higuchi and K. Kitamura, *Nippon Ringaku Kaishi* **35**, 350 (1953).
39. H. Ishikawa, W. J. Schubert, and F. F. Nord, *Life Sciences*, No. 8, p. 365 (1962).
40. H. Ishikawa, W. J. Schubert, and F. F. Nord, *Arch. Biochem. Biophys.* **100**, 131 (1963).

41. H. Ishikawa, W. J. Schubert, and F. F. Nord, *Arch. Biochem. Biophys.* **100**, 140 (1963).
42. H. Ishikawa, W. J. Schubert, and F. F. Nord, *Biochem. Z.* **338**, 153 (1963).
43. D. Keilin and T. Mann, *Nature* **143**, 23 (1939).
44. D. Kertesz and R. Zito, *in* "Oxygenases" (O. Hayaishi, ed.), p. 307. Academic Press, New York, 1962.
45. W. A. Konetzka, M. J. Pelczar, and S. Gottlieb, *J. Bacteriol.* **63**, 771 (1952).
46. K. Kratzl and I. Keller, *Monatsh. Chem.* **197**, 205 (1952).
47. F. Kubowitz, *Biochem. Z.* **292**, 221 (1937); **299**, 32 (1939).
48. K. Law, *Ann. Botany (London)* **14**, 69 (1950); **19**, 561 (1955).
49. L. R. Lawson and C. N. Still, *Tappi* **40**, 56A (1957); "The Biological Decomposition of Lignin." Document Service Center, West Virginia Pulp and Paper Co., Charleston, South Carolina, 1956.
50. B. Leopold, *Acta Chem. Scand.* **4**, 1523 (1950).
51. B. Leopold, *Svensk Kem. Tidskr.* **64**, 1 (1952).
52. B. Leopold, *Svensk Kem. Tidskr.* **64**, 18 (1952).
53. G. Lindberg, *Physiol. Plantarum* **1**, 196 (1948).
54. G. Lindberg, *Nature* **160**, 739 (1950).
55. G. Lindberg and G. Holm, *Physiol. Plantarum* **5**, 100 (1952).
56. B. O. Lindgren and H. Mikawa, *Acta Chem. Scand.* **11**, 836 (1957).
57. Z. Lindroth, *Landwirtsch. Forsch.* **2**, 393 (1904).
58. H. Lyr, *Planta* **49**, 239 (1956).
59. G. Malmström, G. Fåhraeus, and R. Mosbach, *Biochim. Biophys. Acta* **28**, 652 (1958).
60. H. S. Mason, *Advan. Enzymol.* **19**, 79 (1957).
61. H. S. Mason and M. Cronyn, *J. Am. Chem. Soc.* **77**, 491 (1955).
62. R. L. Mitchell and G. J. Ritter, *J. Am. Chem. Soc.* **56**, 1603 (1934).
63. F. F. Nord and J. C. Vitucci, *Advan. Enzymol.* **8**, 253 (1948).
64. M. J. Pelczar, S. Gottlieb, and W. C. Day, *Arch. Biochem.* **25**, 449 (1950).
65. Y. T. Pratt, W. A. Konetzka, M. J. Pelczar, and W. H. Martin, *Appl. Microbiol.* **1**, 171 (1953).
66. J. D. Russel, M. E. K. Henderson, and V. C. Farmer, *Biochim. Biophys. Acta* **52**, 565 (1961).
67. R. L. Schneider, Doctoral Dissertation, Institute of Paper Chemistry, Appleton, Wisconsin, 1961.
68. R. G. H. Siu, "Microbial Decomposition of Cellulose." Reinhold, New York, 1951.
69. D. C. C. Smith, *Nature* **176**, 267, 927 (1955); *J. Chem. Soc.* 2347 (1955).
70. R. Y. Stanier, *J. Bacteriol.* **59**, 527 (1950).
71. J. B. Sumner and G. F. Somers, *in* "Chemistry and Methods of Enzymes," p. 198. Academic Press, New York, 1943.
72. V. Sundman, *Paperi Puu* **43**, 673 (1961).
73. H. H. Tabak, C. W. Chambers, and P. W. Kabler, *J. Bacteriol.* **78**, 469 (1959).
74. W. F. Van Vliet, *Biochim. Biophys. Acta* **15**, 211 (1954).
75. S. A. Waksman and H. W. Smith, *J. Am. Chem. Soc.* **56**, 1225 (1934).
76. C. Wehmer, *Ber. Deut. Botan. Ges.* **45**, 536 (1927).
77. H. Yoshida, *J. Chem. Soc.* **43**, 472 (1883).
78. S. M. Zeller, *Ann. Missouri Botan. Garden* **3**, 439 (1916).
79. W. Zoberst, *Arch. Mikrobiol.* **18**, 1 (1952).

Author Index

Numbers in parentheses are reference numbers and indicate that an author's work is referred to although his name is not cited in the text. Numbers in italic show the page on which the complete reference is listed.

A

Acerbo, S. N., 55(64), 57(2), 59(2), 60(2), 64(2), 66(1, 61, 65), 73, 74
Adkins, H., 28(1, 20), 29(20), 37
Adler, E., 26(2), 37, 53(3), 65(3), 73, 114(2, 3), 115(3), 118
Anderson, A. B. 2(3), 37
Apenitis, A., 89(4), 118
Aulin-Erottman, G., 70(4), 73

B

Ball, C. D., 72(13), 73
Bartlett, V. B., 89(5), 118
Bassett, E. W., 42(1), 51
Bate-Smith, E. C., 63(5), 73
Bavendamm, W., 118
Baylis, P. E. T., 70(6), 73
Bertrand, G., 80(6), 81(9), 82(7), 83(8), 118
Bier, M., 13(54), 15(54), 38
Billek, G., 16(28), 38, 64(46), 66(7), 73, 74
Birkinshaw, J. H., 42(2), 51
Bittner, F., 73
Bjørkman, A., 12(4), 13(5), 37
Björkqvist, K. J., 26(2), 31
Blaisdell, D. J., 80(16), 84(16), 118
Bloom, E. S., 28(1), 37
Blundell, M. J., 21(64), 38, 54(75), 75
Bose, S. R., 80(10), 118
Bourquelot, E., 81(9), 118
Bower, J. R., 28(6), 37
Brauns, D. A., 1(10), 37, 39(4), 45(4), 51, 53(9), 62(8, 9), 63(9). 73, 118

Brauns, F. E.

Brauns, F. E., 1(9, 10), 7(7a, 7b, 8), 8(13), 12(11), 37, 39(3, 4), 45(4), 51, 53(8, 9), 62(8, 9), 63(8, 9), 73, 118
Brewer, C. P., 29(12), 37
Brounstein, C. J., 6(40), 38
Brown, S. A. 54(11), 65(9a), 66(10, 12), 73
Buchanan, M. A., 8(13), 37
Buckland, I. K., 26(14), 37
Burr, G. O., 54(37), 74
Byerrum, R. V., 72(13), 73

C

Calvin, M., 59(14, 55), 73, 74
Campbell, W. A., 80(16), 84(16), 118
Cartwright, K. St. G., 77(13), 79(13), 118
Chambers, C. W., 99(73), 117(73), 119
Cooke, L. M., 28(21), 29(12), 37
Cooper, V. R., 59(70), 75
Coscia, C. J., 30(15), 31(16), 32(16), 37, 67(15), 73
Coulson, C. B., 50(5), 51
Cousin, H., 67(16), 73
Cronyn, M., 71(53), 74, 111(61), 119
Czapek, F., 88(14), 118

D

Dagley, S., 117(15), 118
Davidson, R. W., 80(16), 84(16), 118
Davis, B. D., 40(6, 7, 8), 44(7), 45(26), 51, 57(71), 58(71), 59(71), 60(17, 54), 61(71), 63(18), 73, 74, 75
Dawson, C. R., 81(17), 82(17), 83(17), 118

121

Subject Index

126